I0069837

Math Challenge I-C

Finite Math

Areteem Institute

Math Challenge I-C Finite Math

Series: Math Challenge Curriculum Textbooks, Vol. 21

Edited by   David Reynoso
John Lensmire
Kevin Wang
Kelly Ren

Copyright © 2019 ARETEEM INSTITUTE

WWW.ARETEEM.ORG

PUBLISHED BY ARETEEM PRESS
ALL RIGHTS RESERVED. No part of this publication may be reproduced, stored in a retrieval system, or transmitted, in any form or by any means, electronic, mechanical, photocopying, recording, or otherwise, without prior written permission of the publisher, except for "fair use" or other noncommercial uses as defined in Sections 107 and 108 of the U.S. Copyright Act.

ISBN: 1-944863-40-0
ISBN-13: 978-1-944863-40-1
First printing, February 2019.

TITLES PUBLISHED BY ARETEEM PRESS

Cracking the High School Math Competitions (and Solutions Manual) - Covering AMC 10 & 12, ARML, and ZIML
Mathematical Wisdom in Everyday Life (and Solutions Manual) - From Common Core to Math Competitions
Geometry Problem Solving for Middle School (and Solutions Manual) - From Common Core to Math Competitions
Fun Math Problem Solving For Elementary School (and Solutions Manual)

ZIML MATH COMPETITION BOOK SERIES

ZIML Math Competition Book Division E 2016-2017
ZIML Math Competition Book Division M 2016-2017
ZIML Math Competition Book Division H 2016-2017
ZIML Math Competition Book Jr Varsity 2016-2017
ZIML Math Competition Book Varsity Division 2016-2017
ZIML Math Competition Book Division E 2017-2018
ZIML Math Competition Book Division M 2017-2018
ZIML Math Competition Book Division H 2017-2018
ZIML Math Competition Book Jr Varsity 2017-2018
ZIML Math Competition Book Varsity Division 2017-2018

MATH CHALLENGE CURRICULUM TEXTBOOKS SERIES

Math Challenge I-A Pre-Algebra and Word Problems
Math Challenge I-B Pre-Algebra and Word Problems
Math Challenge I-C Algebra
Math Challenge II-A Algebra
Math Challenge II-B Algebra
Math Challenge III Algebra
Math Challenge I-A Geometry
Math Challenge I-B Geometry
Math Challenge I-C Topics in Algebra
Math Challenge II-A Geometry
Math Challenge II-B Geometry
Math Challenge III Geometry
Math Challenge I-A Counting and Probability
Math Challenge I-B Counting and Probability
Math Challenge I-C Geometry

Math Challenge II-A Combinatorics
Math Challenge II-B Combinatorics
Math Challenge III Combinatorics
Math Challenge I-A Number Theory
Math Challenge I-B Number Theory
Math Challenge I-C Finite Math
Math Challenge II-A Number Theory
Math Challenge II-B Number Theory
Math Challenge III Number Theory

COMING SOON FROM ARETEEM PRESS

Fun Math Problem Solving For Elementary School Vol. 2 (and Solutions Manual)
Counting & Probability for Middle School (and Solutions Manual) - From Common Core to Math Competitions
Number Theory Problem Solving for Middle School (and Solutions Manual) - From Common Core to Math Competitions

The books are available in paperback and eBook formats (including Kindle and other formats).
To order the books, visit `https://areteem.org/bookstore`.

Contents

Copyright © ARETEEM INSTITUTE. All rights reserved.

Introduction

The math challenge curriculum textbook series is designed to help students learn the fundamental mathematical concepts and practice their in-depth problem solving skills with selected exercise problems. Ideally, these textbooks are used together with Areteem Institute's corresponding courses, either taken as live classes or as self-paced classes. According to the experience levels of the students in mathematics, the following courses are offered:

- Fun Math Problem Solving for Elementary School (grades 3-5)
- Algebra Readiness (grade 5; preparing for middle school)
- Math Challenge I-A Series (grades 6-8; intro to problem solving)
- Math Challenge I-B Series (grades 6-8; intro to math contests e.g. AMC 8, ZIML Div M)
- Math Challenge I-C Series (grades 6-8; topics bridging middle and high schools)
- Math Challenge II-A Series (grades 9+ or younger students preparing for AMC 10)
- Math Challenge II-B Series (grades 9+ or younger students preparing for AMC 12)
- Math Challenge III Series (preparing for AIME, ZIML Varsity, or equivalent contests)
- Math Challenge IV Series (Math Olympiad level problem solving)

These courses are designed and developed by educational experts and industry professionals to bring real world applications into the STEM education. These programs are ideal for students who wish to win in Math Competitions (AMC, AIME, USAMO, IMO,

Copyright © ARETEEM INSTITUTE. All rights reserved.

ARML, MathCounts, Math League, Math Olympiad, ZIML, etc.), Science Fairs (County Science Fairs, State Science Fairs, national programs like Intel Science and Engineering Fair, etc.) and Science Olympiad, or purely want to enrich their academic lives by taking more challenges and developing outstanding analytical, logical thinking and creative problem solving skills.

Math Challenge I-C is a four-part course designed to bridge the middle school and high school math materials. For students who participate in the American Math Competitions (AMC), there is a big gap in both the fundamental math concepts and the problem-solving techniques involved between the AMC 8 and AMC 10 contests. This course is developed to help students transition smoothly from middle school to high school, and prepare them for high school math competitions including the AMC 10 & 12, ARML, and ZIML. The full course covers topics and introductory problem solving in algebra, geometry, and finite math. Algebraic topics include linear equations, systems of equations and inequalities, exponents and radicals, factoring polynomials, and solving quadratic equations. Geometric topics include angles in triangles, quadrilaterals, and polygons, congruent and similar polygons, calculating area, and algebraic geometry. Topics in finite math include logic, introductory number theory, and an introduction to probability and statistics. These topics serve as the fundamental knowledge needed for a more advanced problem solving course such as Math Challenge II-A.

The course is divided into four terms:

- Summer, covering Algebra
- Fall, covering covering additional topics in Algebra
- Winter, covering Geometry
- Spring, covering Finite Math

The book contains course materials for Math Challenge I-C: Finite Math.

We recommend that students take all four terms starting with the Summer, but students with the required background are welcome to join for later terms in the course.

Students can sign up for the live or self-paced course at `classes.areteem.org`.

Copyright © ARETEEM INSTITUTE. All rights reserved.

About Areteem Institute

Areteem Institute is an educational institution that develops and provides in-depth and advanced math and science programs for K-12 (Elementary School, Middle School, and High School) students and teachers. Areteem programs are accredited supplementary programs by the Western Association of Schools and Colleges (WASC). Students may attend the Areteem Institute in one or more of the following options:

- Live and real-time face-to-face online classes with audio, video, interactive online whiteboard, and text chatting capabilities;
- Self-paced classes by watching the recordings of the live classes;
- Short video courses for trending math, science, technology, engineering, English, and social studies topics;
- Summer Intensive Camps held on prestigious university campuses and Winter Boot Camps;
- Practice with selected free daily problems and monthly ZIML competitions at ziml.areteem.org.

Areteem courses are designed and developed by educational experts and industry professionals to bring real world applications into STEM education. The programs are ideal for students who wish to build their mathematical strength in order to excel academically and eventually win in Math Competitions (AMC, AIME, USAMO, IMO, ARML, MathCounts, Math Olympiad, ZIML, and other math leagues and tournaments, etc.), Science Fairs (County Science Fairs, State Science Fairs, national programs like Intel Science and Engineering Fair, etc.) and Science Olympiads, or for students who purely want to enrich their academic lives by taking more challenging courses and developing outstanding analytical, logical, and creative problem solving skills.

Since 2004 Areteem Institute has been teaching with methodology that is highly promoted by the new Common Core State Standards: stressing the conceptual level understanding of the math concepts, problem solving techniques, and solving problems with real world applications. With the guidance from experienced and passionate professors, students are motivated to explore concepts deeper by identifying an interesting problem, researching it, analyzing it, and using a critical thinking approach to come up with multiple solutions.

Thousands of math students who have been trained at Areteem have achieved top honors and earned top awards in major national and international math competitions, including Gold Medalists in the International Math Olympiad (IMO), top winners and qualifiers at the USA Math Olympiad (USAMO/JMO) and AIME, top winners at the

Copyright © ARETEEM INSTITUTE. All rights reserved.

Zoom International Math League (ZIML), and top winners at the MathCounts National Competition. Many Areteem Alumni have graduated from high school and gone on to enter their dream colleges such as MIT, Cal Tech, Harvard, Stanford, Yale, Princeton, U Penn, Harvey Mudd College, UC Berkeley, or UCLA. Those who have graduated from colleges are now playing important roles in their fields of endeavor.

Further information about Areteem Institute, as well as updates and errata of this book, can be found online at http://www.areteem.org.

Copyright © ARETEEM INSTITUTE. All rights reserved.

About Zoom International Math League

The Zoom International Math League (ZIML) has a simple goal: provide a platform for students to build and share their passion for math and other STEM fields with students from around the globe. Started in 2008 as the Southern California Mathematical Olympiad, ZIML has a rich history of past participants who have advanced to top tier colleges and prestigious math competitions, including American Math Competitions, MATHCOUNTS, and the International Math Olympaid.

The ZIML Core Online Programs, most available with a free account at ziml.areteem.org, include:

- **Daily Magic Spells:** Provides a problem a day (Monday through Friday) for students to practice, with full solutions available the next day.
- **Weekly Brain Potions:** Provides one problem per week posted in the online discussion forum at ziml.areteem.org. Usually the problem does not have a simple answer, and students can join the discussion to share their thoughts regarding the scenarios described in the problem, explore the math concepts behind the problem, give solutions, and also ask further questions.
- **Monthly Contests:** The ZIML Monthly Contests are held the first weekend of each month during the school year (October through June). Students can compete in one of 5 divisions to test their knowledge and determine their strengths and weaknesses, with winners announced after the competition.
- **Math Competition Practice:** The Practice page contains sample ZIML contests and an archive of AMC-series tests for online practice. The practices simulate the real contest environment with time-limits of the contests automatically controlled by the server.
- **Online Discussion Forum:** The Online Discussion Forum is open for any comments and questions. Other discussions, such as hard Daily Magic Spells or the Weekly Brain Potions are also posted here.

These programs encourage students to participate consistently, so they can track their progress and improvement each year.

In addition to the online programs, ZIML also hosts onsite Local Tournaments and Workshops in various locations in the United States. Each summer, there are onsite ZIML Competitions at held at Areteem Summer Programs, including the National ZIML Convention, which is a two day convention with one day of workshops and one day of competition.

Copyright © ARETEEM INSTITUTE. All rights reserved.

ZIML Monthly Contests are organized into five divisions ranging from upper elementary school to advanced material based on high school math.

- **Varsity:** This is the top division. It covers material on the level of the last 10 questions on the AMC 12 and AIME level. This division is open to all age levels.
- **Junior Varsity:** This is the second highest competition division. It covers material at the AMC 10/12 level and State and National MathCounts level. This division is open to all age levels.
- **Division H:** This division focuses on material from a standard high school curriculum. It covers topics up to and including pre-calculus. This division will serve as excellent practice for students preparing for the math portions of the SAT or ACT. This division is open to all age levels.
- **Division M:** This division focuses on problem solving using math concepts from a standard middle school math curriculum. It covers material at the level of AMC 8 and School or Chapter MathCounts. This division is open to all students who have not started grade 9.
- **Division E:** This division focuses on advanced problem solving with mathematical concepts from upper elementary school. It covers material at a level comparable to MOEMS Division E. This division is open to all students who have not started grade 6.

The ZIML site features are also provided on the ZIML Mobile App, which is available for download from Apple's App Store and Google Play Store.

Acknowledgments

This book contains many years of collaborative work by the staff of Areteem Institute. This book could not have existed without their efforts. Huge thanks go to the Areteem staff for their contributions!

The examples and problems in this book were either created by the Areteem staff or adapted from various sources, including other books and online resources. Especially, some good problems from previous math competitions and contests such as AMC, AIME, ARML, MATHCOUNTS, and ZIML are chosen as examples to illustrate concepts or problem-solving techniques. The original resources are credited whenever possible. However, it is not practical to list all such resources. We extend our gratitude to the original authors of all these resources.

Copyright © ARETEEM INSTITUTE. All rights reserved.

1. Place Values, and Divisibility

Number Theory Introduction

- The theory of numbers goes back 3,000 years or more to the time of the Babylonians.
- The oriental cultures, Hindus and Chinese, also independently developed theories regarding numbers and their properties.
- Number theory, along with geometry, can probably be regarded as the first serious exploration into mathematics.
- Today many problems in math competitions have their roots in number theory.
- More specifically, "number" in Number Theory generally means *whole numbers* or *natural numbers* which include $0, 1, 2, 3, \ldots$. The set of these numbers is often denoted $\mathbb{N}$.
- In general, the real numbers are denoted $\mathbb{R}$. We will sometimes talk about other types of real numbers:
  - Integers: the positive and negative natural numbers, denoted $\mathbb{Z}$.
  - Rationals: the real numbers expressible as a fraction, denoted $\mathbb{Q}$.
  - Irrationals: the real numbers not expressible as a fraction.

Copyright © ARETEEM INSTITUTE. All rights reserved.

Place Values

- The value of a digit depends on its place, or position in a number. For example,

$$654 = 6 \times 100 + 5 \times 10 + 4 \times 1$$
$$2016 = 2 \times 1000 + 0 \times 100 + 1 \times 10 + 6 \times 1$$

- In general,

$$\overline{abc} = a \times 100 + b \times 10 + c$$
$$\overline{a_n a_{n-1} \cdots a_0} = a_n \times 10^n + a_{n-1} \times 10^{n-1} + \cdots + a_0$$

- This can be extended to non-integers as well. For example

$$2.71 = 2 \times 1 + 7 \times \frac{1}{10} + 1 \times \frac{1}{100}.$$

- This system of expressing numbers is referred to as the *decimal* or *base* 10 system. It uses the digits $0, 1, 2, 3, \ldots, 9$.

Divisibility

- Recall if an integer a divides an integer n evenly (that is, there is an integer k such that $a \cdot k = n$), then a is a factor (or divisor) of n. We write $a \mid n$ to denote that a divides n.
- The following basic facts are sometimes useful:
    - If $a \mid n$ and $a \mid m$ then $a \mid n \pm m$.
    - If $a \mid n$ and k is any integer, then $a \mid k \cdot n$.
- If p, q are primes and $p \mid n$ and $q \mid n$ then $p \cdot q \mid n$. We will present the general version of this statement next time.

Divisibility Rules

- The most common divisibility rules are summarized below:

Number	Rule
2	Check whether the last digit is even.
3	Check whether the sum of the digits is divisible by 3.
4	Check whether the last two digits are divisible by 4.
5	Check whether the last digit is 0 or 5.
6	Check whether the number is divisible by 2 and by 3.
9	Check whether the sum of the digits is divisible by 9.
11	Check whether the alternating sum of the digits is divisible by 11.

Copyright © Areteem Institute. All rights reserved.

- Note: The alternating sum alternates adding and subtracting, right to left. For example, the alternating sum of the digits of 1234 is $4 - 3 + 2 - 1 = 2$. Since 2 is not a multiple of 11, 1234 is not divisible by 11.

1.1 Example Questions

Problem 1.1 Formally prove the two basic facts mentioned earlier. Note: These are not hard to prove, but it is useful to see it formally written out.

(a) If $a \mid n$ and $a \mid m$ then $a \mid n \pm m$.

(b) If $a \mid n$ and k is any integer, then $a \mid k \cdot n$.

Problem 1.2 The number 64 has the property that it is divisible by its units digit. How many whole numbers between 10 and 50 have this property?

Problem 1.3 Given a number $\overline{abcde}$. Prove the divisibility rule for:

(a) 5

(b) 9

Problem 1.4 A four digit number $\overline{7a4b}$ is divisible by 18. Find the value of a and b so that this four-digit number has the largest value.

Copyright © ARETEEM INSTITUTE. All rights reserved.

Problem 1.5 Consider the integer $\overline{2a3a1a}$.

(a) If the number is divisible by 9, what are the possible values for a?

(b) If the number is divisible by 11, what are the possible values for a?

Problem 1.6 In the multiplication problem below, A, B, C and D are distinct digits. What is $A + B$?

$$
\begin{array}{r}
A\ B\ A\ B \\
\times\qquad\ \ C\ D \\
\hline
C\ D\ C\ D\ B
\end{array}
$$

Problem 1.7 Suppose you have a 10-digit number made from four 2's, three 3's, two 4's, and one 5. Is it possible for this number to be a perfect square?

Problem 1.8 A five-digit number $\overline{abcde}$ has digits $5,6,7,8,9$ (not necessarily in that order). Assume that $5 \mid \overline{abcde}$, $4 \mid \overline{abcd}$, $3 \mid \overline{abc}$, and $2 \mid \overline{ab}$. Find all possible values of $\overline{abcde}$.

Problem 1.9 A positive integer is equal to 18 times the sum of its digits. What is this number?

Problem 1.10 Show that $\sqrt{3}$ is an irrational number.

Copyright © ARETEEM INSTITUTE. All rights reserved.

1.2 Quick Response Questions

Problem 1.11 Is it true that every decimal number (with a finite number of digits) is a rational number?

Problem 1.12 Is it true that every rational number can be written as a decimal number with a finite number of digits?

Problem 1.13 Which of the following is irrational?

(A) $\dfrac{\sqrt{3}}{\sqrt{27}}$

(B) $\dfrac{4}{13}$

(C) $\dfrac{\sqrt{6}}{\sqrt{2}}$

(D) $-\sqrt{81}$

Problem 1.14 Is the number 5201928 divisible by 9?

Problem 1.15 Is the number 5201928 divisible by 11?

Problem 1.16 What is the smallest number, made up of only 4's (so $4, 44, 444, 4444, \ldots$), that is divisible by 3?

Copyright © ARETEEM INSTITUTE. All rights reserved.

Problem 1.17 Find the number closest to 12345 that is divisible by 9. (If 12345 is divisible by 9, then input 12345.)

Problem 1.18 Find the largest power of 2 (1, 2, 4, 8, etc.) that divides 2121212.

Problem 1.19 Find the smallest number made up of the digits $1, 2, 3, 4$ that is divisible by 11?

Problem 1.20 Find the smallest number consisting of only 1's that is divisible by 9 and 11. How many digits does this number have?

Copyright © ARETEEM INSTITUTE. All rights reserved.

1.3 **Practice Questions**

Problem 1.21 Explain a divisibility rule for 15. Which divisibility rule list in the chart at the beginning of this handout is this rule similar to?

Problem 1.22 The number 24 has the property that it is divisible by its lead digit. How many whole numbers (strictly) between 20 and 60 have this property?

Problem 1.23 Prove the divisibility rule for 11 when $n = \overline{abcd}$.

Problem 1.24 Let $\overline{a357b}$ be a five-digit number. If $44 \mid \overline{a357b}$, find the values of a and b.

Problem 1.25 A four digit number $\overline{3a6b}$ is divisible by 99, find all possible values of a, b.

Problem 1.26 In the addition problem below, A, B, C and D are distinct digits. What is $C + D$?

$$
\begin{array}{cccc}
  & A & B & A \\
+ &   & C & D \\
\hline
  & B & A & B \\
\end{array}
$$

Problem 1.27 Suppose you create an 7-digit number consisting of three 1's and four 3's. Is it possible that your number is a perfect square?

Copyright © Areteem Institute. All rights reserved.

Problem 1.28 A number $\overline{35a2a}$ is divisible by 2. Its last three digits form a three-digit number $\overline{a2a}$ that can be exactly divided by 3. Find all possible such five-digit numbers.

Problem 1.29 An positive integer equals three times the sum of its digits. What is this number?

Problem 1.30 Show that $\sqrt{2}$ is an irrational number.

Copyright © ARETEEM INSTITUTE. All rights reserved.

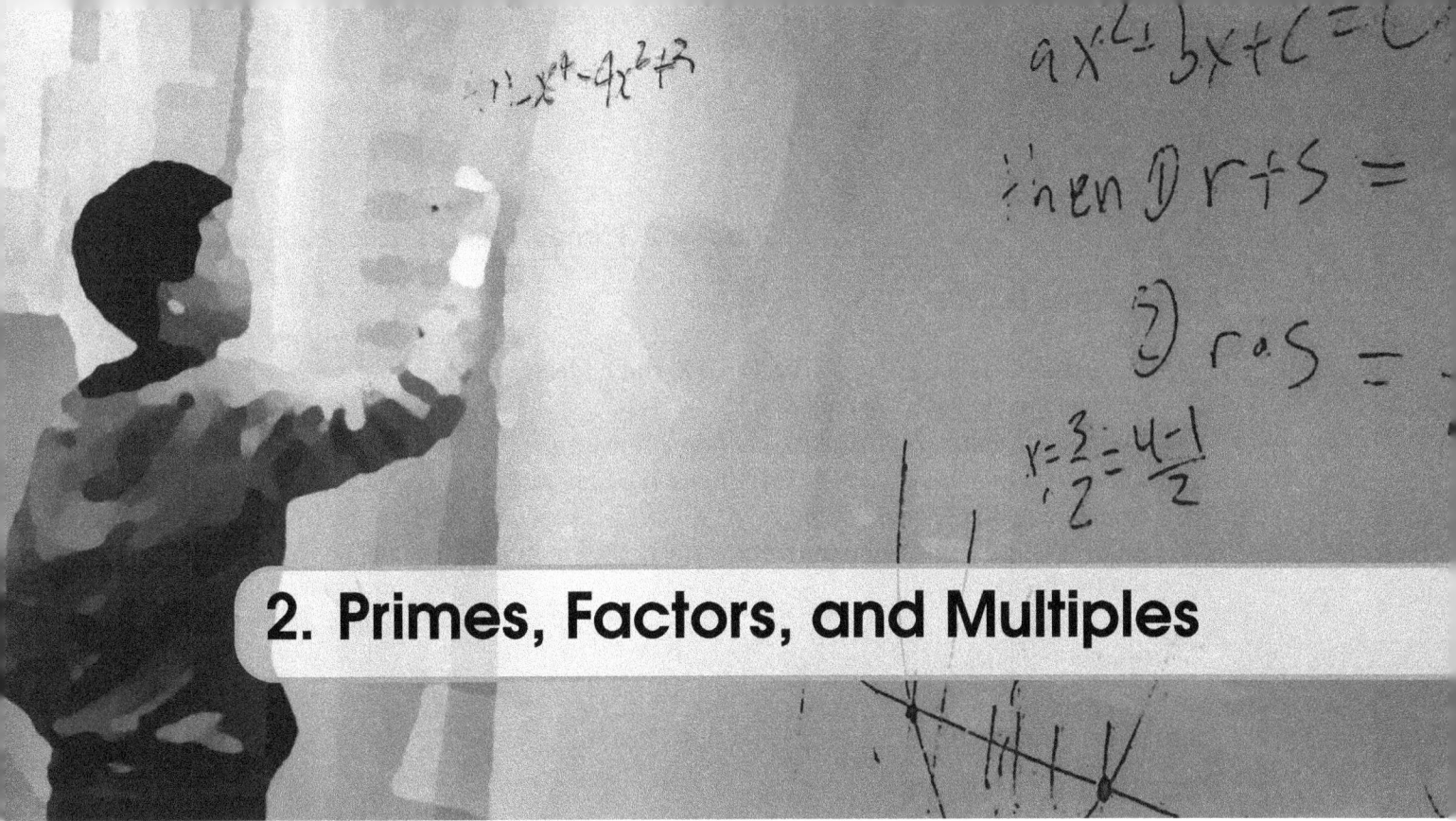

2. Primes, Factors, and Multiples

Review of Divisibility

- Recall if one number a goes evenly into another number n, we say n is divisible by a, denoted $a \mid n$. We say that a is a *divisor* or *factor* of n.
- Equivalently, if $a \mid n$ then there is an integer b such that $n = a \cdot b$. In this case we say n is a multiple of a.
- Therefore, if $a \mid b$, with $n = a \cdot b$, then naturally $b \mid n$ as well. In this sense, the factors/divisors of n can be thought of as coming in "pairs".

Primes and Prime Factorization

- **Euclid**: There are infinitely many primes.
- **Unique Prime Factorization**: Every positive integer n has a unique factorization as a product of primes. That is,

$$n = p_1^{e_1} p_2^{e_2} \cdots p_k^{e_k}$$

  for distinct primes p_i and $e_i > 0$.
- **Divisors**: If $n = p_1^{e_1} p_2^{e_2} \cdots p_k^{e_k}$ and $a \mid n$, then $a = p_1^{f_1} p_2^{f_2} \cdots p_k^{f_k}$ where $0 \leq f_i \leq e_i$.
- **Number of factors**: If $n = p_1^{e_1} p_2^{e_2} \cdots p_k^{e_k}$ is the prime factorization of n, then n has

$$(e_1 + 1)(e_2 + 1) \cdots (e_k + 1)$$

  factors in total. For example, $360 = 2^3 \cdot 2^2 \cdot 5$ has $(3+1)(2+1)(1+1) = 24$ factors.

Copyright © ARETEEM INSTITUTE. All rights reserved.

GCD's and LCM's

- The *greatest common divisor* (GCD) of m and n (denoted $\gcd(m,n)$) is the largest number d such that $d \mid m$ and $d \mid n$.
- We call two numbers m, n *relatively prime* if $\gcd(m,n) = 1$.
- The *least common multiple* (LCM) of m and n (denoted $\operatorname{lcm}(m,n)$) is the smallest number l such that $m \mid l$ and $n \mid l$.
- Recall that we can use the prime factorization of numbers to calculate the GCD and the LCM:
- For example, $42 = 2 \cdot 3 \cdot 7$, $72 = 2^3 \cdot 3^2$, so $\gcd(42,72) = 2 \cdot 3 = 6$ and $\operatorname{lcm}(42,72) = 2^3 \cdot 3^2 \cdot 7 = 504$.

2.1   Example Questions

Problem 2.1 The numerical values of the years are favorite numbers of many math problems. For each of the following, (i) Find the prime factorization and (ii) Find the number of factors.

(a) 2017

(b) 2018

(c) 2019

(d) 2020

Problem 2.2 Prove that a number has an odd number of factors if and only if it is a square.

Copyright © ARETEEM INSTITUTE. All rights reserved.

Problem 2.3 Find the smallest positive integer x such that

(a) $252 \cdot x$ is a perfect square.

(b) $252 \cdot x$ is a perfect cube.

Problem 2.4 Write $10! = A \cdot B \cdot C \cdot D$ for positive integers $A \leq B \leq C \leq D$ with A a factor of B, C, and D. What is the smallest possible value of $D - A$?

Problem 2.5 A natural number is a multiple of 72, and has a total of 15 factors. Find the largest such number.

Problem 2.6 Consider numbers n with the property that the factors of n multiply out to n^3. For example, the factors of 12 are $1, 2, 3, 4, 6, 12$ and

$$1 \cdot 2 \cdot 3 \cdot 4 \cdot 6 \cdot 12 = 1728 = 12^3.$$

In fact, 12 is the smallest such number > 1. What is the next smallest number with this property?

Problem 2.7 Find the greatest common divisor and the least common multiple of

(a) 123 and 1681.

(b) $8!$ and 4^3.

Copyright © ARETEEM INSTITUTE. All rights reserved.

(c) $3! + 5!$ and $5! + 6!$.

Problem 2.8 Consider numbers that leave a remainder of 2 when divided by 3, 4, 5, and 6.

(a) Find the smallest such number.

(b) Find the largest such three-digit number.

Problem 2.9 Suppose A, B, C are integers ≥ 2 with (i) $\gcd(A, B) = 12$, (ii) $\text{lcm}(A, B) = 396$, and (iii) $\gcd(B, C) = 33$. Calculate $\gcd(11A, B)$.

Problem 2.10 Suppose that A has 9 divisors and B has 4 divisors. Find $A + B$ if $\gcd(A, B) = 7$ and $\text{lcm}(A, B) = 2205$.

Copyright © ARETEEM INSTITUTE. All rights reserved.

2.2 Quick Response Questions

Problem 2.11 What is the smallest prime number greater than 50?

Problem 2.12 Two primes are twin primes if one is 2 more than the other. For example, 3 and 5 are a pair of twin primes. How many twin primes pairs are less than 50?

Problem 2.13 Find the smallest prime p such that $2^p - 1$ is not a prime number.

Problem 2.14 Find the number of factors 13005.

Problem 2.15 Find $\gcd(2310, 13005)$.

Problem 2.16 Find $\text{lcm}(2310, 13005)$.

Problem 2.17 Find $\gcd(4200, 3430)$.

Problem 2.18 Find $\text{lcm}(4200, 3430)$.

Problem 2.19 What is the smallest number with 10 factors?

Copyright © ARETEEM INSTITUTE. All rights reserved.

Problem 2.20 The product of all the factors of 2^9 is 2^M for M in integer. What is M?

Copyright © ARETEEM INSTITUTE. All rights reserved.

2.3 Practice Questions

Problem 2.21 Find the prime factorization and number of factors for the following numbers.

(a) 2016

(b) 30030

Problem 2.22 How many numbers from 1 to 1000 (including 1 and 1000) have an odd number of factors?

Problem 2.23 Find the smallest positive integer n such that

(a) $\sqrt{20n}$ is an integer.

(b) $\sqrt{20n}$ is a perfect square.

Problem 2.24 In an earlier question we saw $10! = A \cdot B \cdot C \cdot D = (2^2 \cdot 3)(2^2 \cdot 3 \cdot 5)(2^2 \cdot 3 \cdot 5)(2^2 \cdot 3 \cdot 7)$ where $A \leq B \leq C \leq D$ with A a factor of B, C, and D. If A is not required to be a factor of B, C, and D, in fact $D - A$ can be 8 with $A = 40$ and $D = 48$. What are B and C?

Problem 2.25 The least common multiple of two numbers is 180 and their greatest common divisor is 30. One of the two numbers is 90. what is the other number?

Copyright © ARETEEM INSTITUTE. All rights reserved.

Problem 2.26 Find all numbers n less than 50 with the following property: the product of the divisors is equal to n^3.

Problem 2.27 Find the GCD and LCM of $2! + 3! + 4!$ and $3! + 4! + 5!$.

Problem 2.28 Find the largest three digit integer that has a remainder of 11 when divided by 15, 21, and 35.

Problem 2.29 Suppose A, B, C are integers ≥ 2 with (i) $\gcd(A, B) = 12$, (ii) $\text{lcm}(A, B) = 396$, and (iii) $\gcd(B, C) = 33$.

What are the possibilities for $\gcd(A, C)$?

Problem 2.30 There are 9 divisors for number A and 10 divisors for number B. The least common multiple of A and B is 2800. What are these two numbers?

Copyright © ARETEEM INSTITUTE. All rights reserved.

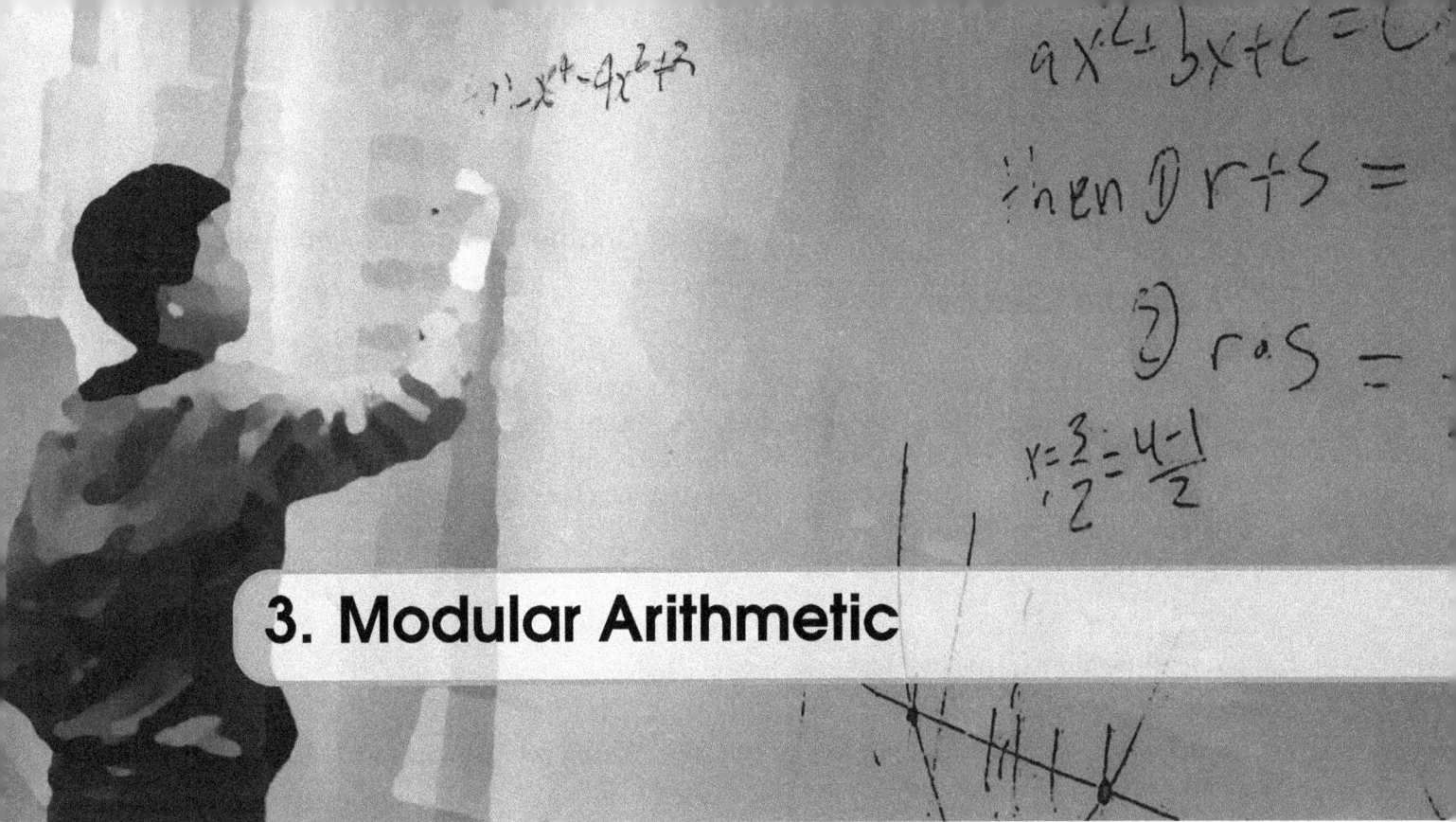

3. Modular Arithmetic

Remainders

- For positive integers a and b, we can always compute $a \div b$, but the result (quotient) might not be an integer.
- We can, however, always write $a = bq + r$, where q is the quotient and r is the remainder, where $0 \le r < b$.
- For example, $38 \div 5 = 7.6$. We can write $38 = 5 \cdot 7 + 3$, where 7 is the quotient and 3 is the remainder.

Modular Arithmetic

- Two numbers a and b are *congruent modulo* m (denoted $a \equiv b \pmod{m}$) if $m \mid (a - b)$.
- Equivalently, a and b have the same remainder when divided by m.
- If we are working modulo m, we often call m the *modulus*.
- In other words, a and b have the same remainder when divided by m.
- For example, there are 7 days per week; use 0 for Sunday, 1 for Monday, ..., 6 for Saturday. Thus, days of the week can be thought of using a modulus of 7. Suppose today is March 14, a Saturday. March 19 is 5 days from now, so it will be $6 + 5 = 11$, and $11 \equiv 4 \pmod{7}$, thus it is Thursday.
- The "Clock Arithmetic" is also an example of modular arithmetic. The modulus is 12. Suppose it is 11 o'clock now. 6 hours from now, $11 + 6 = 17 \equiv 5 \pmod{12}$, so it will be 5 o'clock (switched am/pm). If we want to calculate based on 24-hour time, then use modulo 24.

Copyright © ARETEEM INSTITUTE. All rights reserved.

3.1 Example Questions

Problem 3.1 A group of pirates went to hunt for treasure. They found a chest of gold coins. They tried to equally divide the coins, but 5 coins were left over. So they picked one pirate among themselves and threw him overboard. Then they tried to divide the coins again, but now 10 coins were left over! If the chest held 100 coins, how many pirates were there originally?

Problem 3.2 Everyday Problems with Remainders

(a) Suppose that the date is Saturday March 26th. What day of the week will March 6th be next year? (Assume next year is not a leap year.)

(b) Suppose it is 9 o'clock now. What time will it be 100 hours from now, if we ignore am/pm?

(c) What might be a better way to think of a $1000°$ angle?

Problem 3.3 Patterns!

(a) Find the units digit of 2^{2018}.

(b) Find the remainder when 2^{2018} is divided by 9.

Problem 3.4 Prove the equivalence mentioned in the beginning of the packet: $m \mid (a-b)$ if and only if a and b have the same remainder when divided by m.

Copyright © ARETEEM INSTITUTE. All rights reserved.

Problem 3.5 Assume that $a \equiv b \pmod{m}$ and $c \equiv d \pmod{m}$. Are the following true or false? If false, come up with a counterexample. If true, you'll prove it on your homework!

(a) If $b \equiv c \pmod{m}$ then $a \equiv c \pmod{m}$.

(b) $(a + c) \equiv (b + d) \pmod{m}$.

(c) $(a \cdot c) \equiv (b \cdot d) \pmod{m}$.

(d) If k is an integer and $k \mid a, k \mid b$, then $(a/k) \equiv (b/k) \pmod{m}$.

(e) If n is a positive integer, then $a^n \equiv b^n$.

Problem 3.6 Find the remainder when

(a) $35^{53} + 53^{35}$ is divided by 10.

(b) $31^{2018} + 33^{2018}$ is divided by 32.

Problem 3.7 If $m > 1$ and $60 \equiv 70 \equiv 85 \pmod{m}$, what is m?

Problem 3.8 Suppose A, B, C, D are 4 consecutive natural numbers.

Copyright © ARETEEM INSTITUTE. All rights reserved.

(a) Find the remainder when $A+B+C+D$ is divided by 4.

(b) Suppose you also know that $A+B+C+D$ is a three-digit number between 400 and 440 and $A+B+C+D$ is divisible by 9. Find A,B,C,D.

Problem 3.9 Let $n = \overline{a_k a_{k-1} \ldots a_1 a_0} = a_k 10^k + a_{k-1} 10^{k-1} + \cdots + a_1 10 + a_0, a_i \in \{0,1,\ldots,9\}$. Prove the following. Note: These are similar to things you've already proven, but practice using modular arithmetic here!

(a) Prove that $n \equiv \overline{a_{j-1} a_{j-2} \ldots a_1 a_0} \pmod{2^j}$.

(b) Prove that $n \equiv (a_k + a_{k-1} + \cdots + a_1 + a_0) \pmod 9$. Note this means that a number is equal to the sum of its digits modulo 9.

Problem 3.10 Concatenate the positive integers $1, 2, 3, \ldots, 2017$ to form a new integer:

$$12345678910111213142 \cdots 201520162017.$$

What is the remainder when this new integer is divided by 9?

Copyright © ARETEEM INSTITUTE. All rights reserved.

3.2 Quick Response Questions

Problem 3.11 A certain store is running a promotion that starts at 12 AM on Friday. If the promotion runs for 80 hours, what hour will it be when the promotion ends? (Ignore AM or PM in your answer.)

Problem 3.12 At the end of the year my school has a talent show. Each participant has exactly 10 minutes to do their presentation. Mr. Tahoe arrived late, and only stayed for 30 minutes. What is the maximum number of participants Mr. Tahoe would be able to see on stage?

Problem 3.13 Are 345 and 543 congruent modulo 13?

Problem 3.14 Are 41 and -56 congruent modulo 97?

Problem 3.15 Find the last two digits of 216×348.

Problem 3.16 Find the remainder when 51374948 is divided by 9.

Problem 3.17 What is the units digit of 2^{827}?

Problem 3.18 What is the remainder of 576^2 upon division by 10?

Copyright © ARETEEM INSTITUTE. All rights reserved.

Problem 3.19 What is the units digit of $19^{19} + 99^{99}$?

Problem 3.20 What is the remainder of $1! + 2! + 3! + \cdots + 2016!$ upon division by 5?

Copyright © ARETEEM INSTITUTE. All rights reserved.

3.3   Practice Questions

Problem 3.21 A group of pirates went to hunt for treasure. They found a chest of gold coins. They tried to equally divide the coins, but 5 coins were left over. So they picked one pirate among themselves and threw him overboard. Then they tried to divide the coins again, but now 10 coins were left over! So again they picked another pirate among themselves and threw him overboard and tried to divide the coins again. This process continued until the coins could be evenly divided. If the chest held 100 coins, how many coins did each pirate get when this process was completed?

Problem 3.22 David is afraid of Friday the 13th, particularly when it is in April. This year there is a Friday the 13th in April. In how many years will there be again a Friday the 13th in April? (Remember a normal year has 365 days and a leap year has one extra day. Leap years occur every 4 years and 2016 was a leap year.)

Problem 3.23 What are the last two digits of 91^{91}?

Problem 3.24 Using the fact that $a \equiv b \pmod{m}$ is equivalent to $m \mid a - b$, prove that if $a \equiv b \pmod{m}$ and $b \equiv c \pmod{m}$, then $a \equiv c \pmod{m}$.

Problem 3.25 Assume that $a \equiv b \pmod{m}$ and $c \equiv d \pmod{m}$. Prove that

(a) $(a + c) \equiv (b + d) \pmod{m}$.

(b) $(a \cdot c) \equiv (b \cdot d) \pmod{m}$.

Copyright © ARETEEM INSTITUTE. All rights reserved.

Problem 3.26 Find the remainder when $2016^{2017} + 2020^{2017}$ is divided by 2018.

Problem 3.27 For how many $m > 1$ is it true that $1023 \equiv 981 \pmod{m}$?

Problem 3.28 The Fibonacci sequence is defined by $F_1 = F_2 = 1$, and $F_{n+2} = F_{n+1} + F_n$, that is, the first two terms are both 1, and each subsequent term is the sum of the previous two terms. Find the remainder when F_{2018} is divided by 11.

Problem 3.29 Let $n = \overline{abcdef} = a \cdot 10^5 + b \cdot 10^4 + \cdots + e \cdot 10^1 + f$. Prove that $n \equiv f - e + d - c + b - a \pmod{11}$.

Problem 3.30 Find the remainder when the sum $n = 2 + 212 + 21212 + \cdots + 21212121212$ (the last term has five 21's and a 2) is divided by 11.

Copyright © ARETEEM INSTITUTE. All rights reserved.

4. Sequences and Series

Sequences

- A *sequence* is a list of numbers (either finite or infinite) $a_0, a_1, a_2, \ldots$.
- For example:
    - $1, 2, 3, 4, 3, 2, 1$ is a finite sequence, with *length* 7.
    - $1, 2, 4, 8, 16, \ldots$ is an infinite sequence, where we have a formula: $a_n = 2^n$. Remember we start with a_0.
    - $2, 3, 5, 7, 11, \ldots$ is an infinite sequence listing all the prime numbers.
- As we've seen in the examples, not all sequences need formulas! In fact, it might be very hard to come up with formulas for some sequences.

Arithmetic and Geometric Sequences

- An *arithmetic* sequence has formula $a_n = a_0 + k \cdot n$ where a_0 is the starting value and k is the common difference.
- For example, $2, 5, 8, 11, \ldots$ is an arithmetic sequence with formula $a_n = 2 + 3n$.
- An *geometric* sequence has formula $a_n = a_0 \cdot r^n$ where a_0 is the starting value and r is the common ratio.
- For example, $3, 6, 12, 24, \ldots$ is a geometric sequence with formula $a_n = 3 \cdot 2^n$.

Copyright © Areteem Institute. All rights reserved.

Differences

- Given a sequence, the difference of the sequence is the sequence of differences of the terms in the sequence.
- For example, the sequence $1, 3, 5, 7, 9, \ldots$ has differences $2, 2, 2, 2, \ldots$.
- In general the sequence $a_0, a_1, a_2, \ldots$ has sequence of differences $a_1 - a_0, a_2 - a_1, a_3, -a_2, \ldots$.
- If this process is repeated, we call them the first differences, second differences, etc.
- **Fact:** If a sequence has an equation that is a polynomial of degree n, then the $(n+1)$st differences are all 0.

Series and Summation Notation

- We call the sum (whether finite or infinite) of a sequence a *series*.
- For convenience, we often use *summation notation* when writing sequences, where

$$a_j + a_{j+1} + a_{j+2} + \cdots + a_k = \sum_{n=j}^{k} a_n.$$

- For example,

  ○ $1 + 2 + 3 + 4 + \cdots + n = \sum_{k=1}^{n} k$

  ○ $1 + 2 + 4 + 8 + 16 + \cdots + 256 = \sum_{k=0}^{8} 2^k$

  ○ $1 + \dfrac{1}{2} + \dfrac{1}{4} + \dfrac{1}{8} + \cdots = \sum_{k=0}^{\infty} \left(\dfrac{1}{2}\right)^k$

4.1 Example Questions

Problem 4.1 For each of the following descriptions of a sequence: (i) write out the first few terms of the sequence, (ii) state whether it is an infinite or finite sequence, (iii) if it is a finite sequence, state its length.

(a) The number of days in the month, starting with January, February, etc.

Copyright © ARETEEM INSTITUTE. All rights reserved.

(b) The sequence with where the nth term is $4n^2 + 2$.

(c) The sequence where the nth term is the remainder when 2^n is divided by 9.

Problem 4.2 Arithmetic Sequences

(a) A sequence is given with the formula $a_n = 8 - 7n$. Write out the first 5 terms of this sequence.

(b) A sequence starts with the terms $-4, 8, 20, 32, 44, \ldots$. Find a general equation for a_n. Remember $a_0 = -4$.

Problem 4.3 Geometric Sequences

(a) A sequence is given with the formula $a_n = 2 \cdot (-2)^n$. Write out the first 5 terms of this sequence.

(b) A sequence starts with the terms $625, 125, 25, 5, 1, \ldots$. Find a general equation for a_n. Remember $a_0 = 625$.

Problem 4.4 For each of the following sequences, find the first few differences. Do you notice any patterns?

(a) Arithmetic Sequence: $a_n = 3 + 5n$

Copyright © ARETEEM INSTITUTE. All rights reserved.

(b) Geometric Sequence: $a_n = 2 \cdot 2^n$

(c) Cubic Sequence: $a_n = n^3 + 1$

Problem 4.5 Consider the sequence $2, 0, 0, 2, 6, 12, 20, \ldots$. Use the method of differences, prove that this sequence has a quadratic equation and find the equation.

Problem 4.6 Summation Notation Introduction

(a) Write out the terms of the series $\sum\limits_{n=2}^{6} 3n^2$ and calculate the sum.

(b) Write out the terms of the series $\sum\limits_{n=1}^{5} (-1)^n \dfrac{1}{n}$ and calculate the sum.

(c) Find the sum $2 + 9 + 28 + 65 + 126$. What is this in summation notation?

(d) Find the sum $-3 + 9 - 27 + 81 - 243 + 729 - 2187$. What is this in summation notation?

Problem 4.7 Arithmetic Series

(a) Calculate $\sum\limits_{k=1}^{n} k$.

Copyright © ARETEEM INSTITUTE. All rights reserved.

(b) Calculate the sum of the first 20 terms of the sequence $7, 5, 3, 1, -1, \ldots$.

(c) Suppose $a_n = 3n - 2$. Calculate $\displaystyle\sum_{k=2}^{18} a_k$.

Problem 4.8 The repeating decimal $0.\overline{9}$

(a) Write the repeating fraction $0.\overline{9}$ as an (infinite) geometric series.

(b) Let $0.\overline{9} = x$. Consider the expression $10x - x$ and use this expression to solve for x.

Problem 4.9 Geometric Series

(a) Calculate the sums $\displaystyle\sum_{k=0}^{n} 2^k$ for $n = 1, 2, 3, 4, 5$. Do you notice a pattern?

(b) Using a method similar to the previous problem, explain a general formula for $\displaystyle\sum_{k=0}^{n} 2^k$.

(c) For what values of $r \neq \pm 1$ in a geometric sequence $a_n = a_0 \cdot r^n$ does it make sense to have the infinite series $\displaystyle\sum_{k=0}^{\infty} a_k$?

Copyright © ARETEEM INSTITUTE. All rights reserved.

(d) Calculate $\displaystyle\sum_{k=0}^{\infty} 9 \cdot \left(\frac{1}{3}\right)^{k}$.

Problem 4.10 What is the units digit of $1^2 + 2^2 + 3^2 + \cdots + 99^2$?

Copyright © Areteem Institute. All rights reserved.

4.2 **Quick Response Questions**

Problem 4.11 What is the next term in the sequence $0, 3, 8, 15, 24, \ldots$?

Problem 4.12 Which of the following is a formula for the sequence $3, -3, -9, -15, -21, \ldots$?

(A) $a_n = -3 - 6 \cdot n$
(B) $a_n = 3 - 6 \cdot n$
(C) $a_n = (3 - 6) \cdot n$
(D) $a_n = -3 + 6 \cdot n$

Problem 4.13 What is the sum of the first 15 terms of the sequence $a_n = (-1)^n (n + 1)$?

Problem 4.14 What is the 10th term of the sequence $-3, 6, -12, 24, -48, \ldots$?

Problem 4.15 Which of the following are the second differences of the sequence $4, 12, 36, 108, 324, 972, \ldots$?

(A) $4, 12, 36, 108, 324, \ldots$
(B) $8, 24, 72, 216, 648, \ldots$
(C) $12, 36, 108, 324, 972, \ldots$
(D) $16, 48, 144, 432, 1296, \ldots$

Copyright © ARETEEM INSTITUTE. All rights reserved.

Problem 4.16 Which of the following represents the sum $-4 + 9 - 16 + 25 - \cdots - 100 + 121$ using summation notation?

(A) $\sum_{n=2}^{11} -n^2$

(B) $\sum_{n=1}^{10} (-n+1)^2$

(C) $\sum_{n=2}^{11} (-1)^{n+1} n^2$

(D) $\sum_{n=2}^{11} (-n)^2$

Problem 4.17 Calculate $\displaystyle\sum_{n=1}^{100} n$

Problem 4.18 The number $1.\overline{8}$ is equal to $\dfrac{P}{Q}$ as a reduced fraction. What is $Q - P$?

Problem 4.19 Calculate $\displaystyle\sum_{k=0}^{\infty} 2 \cdot \left(\frac{1}{2}\right)^k$.

Problem 4.20 Calculate $\displaystyle\sum_{k=0}^{10} 2^k$?

Copyright © ARETEEM INSTITUTE. All rights reserved.

4.3 Practice Questions

Problem 4.21 Consider the sequence where a_n equals the remainder when $n^2 + 3$ is divided by 5.

(a) Write out the first 10 terms of the sequence.

(b) The sequence is an infinite sequence, but is has a cycle of length N. What is N? Can you explain the value of N using modular arithmetic?

Problem 4.22 Find a formula for the following arithmetic sequences.

(a) The arithmetic sequence which starts $-3, 1, 5, 9, \ldots$.

(b) The arithmetic sequence with $a_3 = 5$ and $a_6 = -13$.

Problem 4.23 Find a formula for the following geometric sequences.

(a) The geometric sequence which starts $27, 18, 12, 8, \ldots$.

(b) The geometric sequence with $a_3 = -1$ and $a_7 = -256$ if $a_0 > 0$.

Problem 4.24 Differences Further Practice

Copyright © ARETEEM INSTITUTE. All rights reserved.

(a) Calculate the first few terms of the first, second, and third differences for the sequence $a_n = -n^2 + 3n - 8$. Verify that the third differences you calculate are all 0.

(b) Calculate the first few terms of the first and second differences for the sequence $a_n = 3^n$. Do you notice any patterns?

Problem 4.25 Let $a_n = 0^2 + 1^2 + 2^2 + \cdots + n^2$ so a_n starts $0, 1, 5, 14, 30, 55, \ldots$. Using the method of differences, argue that there exists a cubic formula for a_n. Note: You do not need to actually find the formula, but you can as a challenge!

Problem 4.26 One of the reasons summation notation is confusing is that certain series can be written in multiple ways. For example

$$\sum_{k=1}^{5} 2k = \sum_{k=0}^{4} 2k + 2.$$

For each of the following sequences, rewrite them so they start with $k = 0$ as in the example above.

(a) $\sum_{k=1}^{10} 2 \cdot 2^k$.

(b) $\sum_{k=3}^{18} 20 - 4k$.

Copyright © Areteem Institute. All rights reserved.

Problem 4.27 Find a general formula for a finite arithmetic series $\sum_{k=0}^{n-1} a_k = a_0 + a_1 + \cdots + a_{n-1}$. (Here the sum is for n terms total.)

Problem 4.28 Convert the decimal $0.\overline{78}$ into a fraction.

Problem 4.29 Geometric Series

(a) Find a general formula for a finite geometric series $\sum_{k=0}^{n-1} a_0 \cdot r^k = a + ar + ar^2 + \cdots ar^{n-1}$. (Here the sum is for n terms total.)

(b) For $0 < r < 1$, find a general formula for an infinite geometric series $\sum_{k=0}^{\infty} a_0 \cdot r^k = a + ar + ar^2 + \cdots$.

Problem 4.30 What is the remainder when $0^2 + 1^2 + 2^2 + 3^2 + \cdots + 99^2$ is divided by 9?

Copyright © AReteem Institute. All rights reserved.

5. Recursive Sequences

Recursive Formulas

- In the sequence $3, 7, 15, 31, 63, \ldots$, each term is twice the previous term plus one and the first term is 3. In symbols, $a_{n+1} = 2 \cdot a_n + 1$, with $a_0 = 3$.
- This type of description of a sequence is called a *recursive* formula, where the sequence is defined using the previous terms in the sequence. Note in a recursive formula you may need to specify the first term (or first few terms) of the sequence.
- For example, the sequence $0, 1, 2, 5, 12, 29, \ldots$ can be described by saying $a_0 = 0, a_1 = 1$ and $a_{n+1} = 2 \cdot a_n + a_{n-1}$. $(2 = 2 \cdot 1 + 0, 5 = 2 \cdot 2 + 1$, etc.)

Arithmetic and Geometric Sequences

- An *arithmetic* sequence has formula $a_n = a_0 + k \cdot n$ where a_0 is the starting value and k is the common difference.
- Alternatively, an arithmetic sequence has recursive formula $a_{n+1} = a_n + k$.
- For example, $2, 5, 8, 11, \ldots$ is an arithmetic sequence with formula $a_n = 2 + 3n$ and recursive formula $a_0 = 2, a_{n+1} = a_n + 3$.
- An *geometric* sequence has formula $a_n = a_0 \cdot r^n$ where a_0 is the starting value and r is the common ratio.
- Alternatively, a geometric sequence has recursive formula $a_{n+1} = r \cdot a_n$.
- For example, $3, 6, 12, 24, \ldots$ is a geometric sequence with formula $a_n = 3 \cdot 2^n$ and recursive formula $a_0 = 3, a_{n+1} = 2 \cdot a_n$.

Copyright © ARETEEM INSTITUTE. All rights reserved.

5.1 Example Questions

Problem 5.1 List the first few terms of each sequence and verify that both formulas lead to the same sequence:

$a_n = 5 + 3n$ and $b_0 = 5$, $b_{n+1} = b_n + 3$

Problem 5.2 List the first few terms of each sequence and verify that both formulas lead to the same sequence:

$a_n = 2 \cdot 3^n$ and $b_0 = 2$, $b_{n+1} = b_n \cdot 3$

Problem 5.3 Arithmetic sequences are given below, in one of three ways: (i) the first few terms of the sequence, (ii) the formula, or (iii) the recursive formula. Give the other 2 ways of describing the sequence. (That is, if the recursive formula is given, write out the first few terms and give the general formula for the sequence.)

(a) $a_0 = 5$, $a_{n+1} = 8 + a_n$.

(b) $3, 5, 7, \ldots$.

(c) $a_n = 6 - 5n$.

Problem 5.4 Geometric sequences are given below, in one of three ways: (i) the first few terms of the sequence, (ii) the formula, or (iii) the recursive formula. Give the other 2 ways of describing the sequence. (That is, if the recursive formula is given, write out the first few terms and give the general formula for the sequence.)

Copyright © ARETEEM INSTITUTE. All rights reserved.

(a) $2, 4, 8, \ldots$

(b) $a_0 = -3$, $a_{n+1} = -2 \cdot a_n$.

(c) $a_n = 4^n$.

Problem 5.5 Verify algebraically that the formulas and recursive formulas lead to the same sequence
(a) $a_n = 5 + 4n$; $a_0 = 5$, $a_{n+1} = a_n + 4$.

(b) $a_n = 3 \cdot 2^n$; $a_0 = 3$, $a_{n+1} = 2 \cdot a_n$.

Problem 5.6 Suppose a sequence starts $G_0 = 2$, $G_1 = 1$, $G_{n+1} = 2 \times G_n - G_{n-1}$. That is, multiply the previous term by 2 and subtract the term before that. Is there a simpler formula for this sequence?

Problem 5.7 Review of common differences
(a) Use common differences to determine what kind of formula the recursive sequence $a_3 = 0$, $a_{n+1} = a_n + n - 1$ has.

(b) Find a formula for the sequence.

Copyright © ARETEEM INSTITUTE. All rights reserved.

Problem 5.8 Find a formula for the sequence defined by the recursive formula $a_1 = 2$, $a_{n+1} = 3a_n + 2$

Problem 5.9 Recall the Fibonacci sequence defined by $F_1 = F_2 = 1$, and $F_{n+2} = F_{n+1} + F_n$. For $n \geq 1$, let S_n be the sum of the first n terms of the Fibonacci sequence. ($S_n = F_1 + \cdots + F_n$ or $S_1 = F_1$, $S_2 = F_1 + F_2$, etc.)

(a) Write out S_n for $n = 1, 2, 3, 4, 5$.

(b) Compare your answer in (a) with the original Fibonacci sequence.

Problem 5.10 More on recursive sequences

(a) Define a sequence generated by the following: start with 12 and divide by 2 if the number is even or take 3 times the number plus 1 if the number is odd. What is the largest value a term of this sequence can have?

(b) What is the 8th term in the sequence 1, 11, 21, 1211, 111221, 312211?

Copyright © ARETEEM INSTITUTE. All rights reserved.

5.2 Quick Response Questions

Problem 5.11 Suppose a sequence has recursive formula $a_0 = 2, a_{n+1} = 3a_n + n$. What is a_4?

Problem 5.12 Suppose a sequence has recursive formula $a_0 = 1, a_1 = 0$ and $a_{n+1} = n \cdot (a_n + a_{n-1})$. What is a_4?

Problem 5.13 Which of the following is a recursive definition for the sequence $a_n = 50 - 4n$?

(A) $a_0 = 46, a_{n+1} = a_n - 4$
(B) $a_0 = 46, a_{n+1} = 4 - a_n$
(C) $a_0 = 50, a_{n+1} = a_n - 4$
(D) $a_0 = 50, a_{n+1} = -4 \cdot a_n$

Problem 5.14 Find a recursive definition for the sequence $a_n = -3 \cdot (-2)^n$.

(A) $a_0 = 3, a_{n+1} = a_n - 2$
(B) $a_0 = -3, a_{n+1} = a_n - 2$
(C) $a_0 = 3, a_{n+1} = -2 \cdot a_n$
(D) $a_0 = -3, a_{n+1} = -2 \cdot a_n$

Problem 5.15 Suppose a sequence has recursive definition $a_0 = 200, a_{n+1} = a_n - 3$. Find a_{100}.

Problem 5.16 Suppose a sequence has recursive definition $a_0 = \dfrac{1}{27}, a_{n+1} = 3a_n$. Find a_8.

Copyright © ARETEEM INSTITUTE. All rights reserved.

Problem 5.17 Write out the first few terms of the recurrence relation $a_0 = 2, a_{n+1} = \dfrac{1}{1 - a_n}$ until you find a pattern. This pattern repeats every K terms. What is K?

Problem 5.18 Which of the options below gives a general formula for a sequence with recursive definition $a_0 = 4, a_{n+1} = 3 \cdot a_n - 2$?

(A) $a_n = 3^n + 3$
(B) $a_n = 3^{n+1} + 3$
(C) $a_n = 6n + 4$
(D) $a_n = 3 \cdot 3^n + 1$

Problem 5.19 The Lucas Sequence is defined similarly as the Fibonacci Sequence except the sequence begins with the first term 2 and second term 1. Therefore the Lucas Sequence starts $2, 1, 3, 4, 7, 11$. What is the first number in the Lucas Sequence that is larger than 50?

Problem 5.20 Let F_n denote the Fibonacci sequence and L_n denote the Lucas sequence. ($F_0 = 0, F_1 = 1$ and $L_0 = 2, L_1 = 1$ with each term thereafter is the sum of the previous two number in the sequence.) Which of the following is not an expression of L_n in terms of the Fibonacci sequence?

(A) $L_n = F_{n+2} - F_{n-2}$
(B) $L_n = F_{n+1} - F_{n-1}$
(C) $L_n = F_{n+1} + F_{n-1}$
(D) $L_n = F_n + 2F_{n-1}$

Copyright © Areteem Institute. All rights reserved.

5.3 Practice Questions

Problem 5.21 List the first few terms of the sequence $a_0 = 0$ and $a_{n+1} = a_n + (n+1)^2$. Verify that these satisfy the formula $a_n = \dfrac{n(n+1)(2n+1)}{6}$.

Problem 5.22 List the first few terms of the sequence $a_0 = 0$, $a_1 = 1$, and $a_{n+1} = 2a_n - a_{n-1} + 3$. Verify that these satisfy the formula $\dfrac{n(3n-1)}{2}$.

Problem 5.23 Describe a pattern to determine the numbers in the following sequence:

$$-3, 10, 23, 36, \ldots$$

Give both a recursive formula and a general formula.

Problem 5.24 Describe a pattern to determine the numbers in the following sequence:

$$16, 24, 36, 54, 81, \ldots$$

Problem 5.25 Consider a sequence with recursive formula $a_0 = 2$ and $a_{n+1} = 2a_n - (3 \cdot (-1)^n)$.

Verify algebraically that this sequence satisfies the formula $a_n = 2^n + (-1)^n$.

Copyright © ARETEEM INSTITUTE. All rights reserved.

Problem 5.26 A sequence has recursive definition $H_0 = 1$, $H_1 = -1$, and $H_{n+1} = -2H_n - H_{n-1}$. Write out a few terms of this sequence and come up with a formula for H_n.

Problem 5.27 Consider a (convex) polygon with n sides (with $n \geq 3$). Let d_n denote the number of diagonals the polygon has. (So $d_3 = 0$, $d_4 = 2$, etc.) Show that d_n satisfies the recursive formula $d_3 = 0$ and $d_{n+1} = d_n + n - 1$.

Note this implies that $d_n = \dfrac{n(n-3)}{2}$ by the problem from earlier in the handout.

Problem 5.28 Find a formula for the sequence defined by the recursive formula $a_0 = 5$ and $a_{n+1} = 2a_n - 3$.

Problem 5.29 Let F_n denote the nth Fibonacci number with $F_1 = F_2 = 1$. Let S_n be the sum of the squares of the first n terms of the Fibonacci sequence. ($S_1 = F_1^2$, $S_2 = F_1^2 + F_2^2$, etc.) Let $T_n = F_n \cdot F_{n+1}$. Write out the first few terms of S_n and T_n. What pattern do you notice?

Problem 5.30 A recursive sequence is defined as follows. If the current number if x, the next number is $x \div 2$ if x is even and $3x + 1$ if x is odd.

Try out this sequence starting with a few different positive integers. Do you notice any pattern? Make a guess about the behavior of this sequence.

Copyright © ARETEEM INSTITUTE. All rights reserved.

6. Counting Introduction

Counting by Enumeration

- It is a natural tendency for us to count using our fingers: $1, 2, 3, \ldots$.
- To count how many ways something can happen, we can list out all the possibilities, and then count them up. We will call this method enumeration, a fancy word for "counting."
- It will be useful to organize the way we enumerate the outcomes to avoid mistakes. Further, enumeration can often help us find patterns we may not have noticed when we first started the problem.

Sum and Product Rule

- **Sequential Counting Principle (Product Rule)**: Suppose that a procedure can be broken down into k successive tasks. If there are n_1 ways to do the first task, and n_2 ways to do the second task after the first task has been done, and so on, then there are $n_1 \times n_2 \times \cdots \times n_k$ ways to do the procedure.
- **Additive Counting Principle (Sum Rule)**: Suppose we have tasks $T_1, T_2, \ldots, T_k$ that can be done in $n_1, n_2, \ldots, n_k$ ways, respectively, and no two of these tasks can be done at the same time, then there are $n_1 + n_2 + \ldots + n_k$ ways to do one of these tasks.
- **Note**: Both of these rules are "reversible". For example, suppose a procedure can be broken down into 2 successive tasks. If there are n ways to do the entire procedure and m ways to do the 2nd task, then there are $\dfrac{n}{m}$ ways of doing the 1st task.

Copyright © ARETEEM INSTITUTE. All rights reserved.

Permutations and Combinations

- **Permutations**: Permutation means arrangement of things *in a certain order*. The number of permutations of r elements taken out of a set of n elements (without repeating) is denoted $_nP_r$:

$$_nP_r = n(n-1)(n-2)\cdots(n-r+1) = \frac{n!}{(n-r)!}.$$

- **Combinations**: Combination means selection of things where *order does not matter*. The number of combinations of r elements taken out of a set of n elements is denoted $_nC_r$ or $\binom{n}{r}$:

$$_nC_r = \binom{n}{r} = \frac{n(n-1)(n-2)\cdots(n-r+1)}{r!} = \frac{n!}{r!(n-r)!}.$$

- **Relationship with Pascal's Triangle**: Here is Pascal's triangle, written both in the usual way, and written with its terms expressed as combinations.

$$\binom{0}{0}$$

$$\binom{1}{0} \qquad \binom{1}{1}$$

$$\binom{2}{0} \qquad \binom{2}{1} \qquad \binom{2}{2}$$

$$\binom{3}{0} \qquad \binom{3}{1} \qquad \binom{3}{2} \qquad \binom{3}{3}$$

$$1$$
$$1 \qquad 1$$
$$1 \qquad 2 \qquad 1$$
$$1 \qquad 3 \qquad 3 \qquad 1$$

In other words, the entries in Pascal's triangle equal to the corresponding entry in the triangle of combination coefficients on the left. We'll explore this in more detail later on in the course.

6.1 Example Questions

Problem 6.1 Suppose John has 2 hats, 5 shirts, 1 jacket, 4 pairs of pants, 3 pairs of shorts, and 4 pairs of shoes.

(a) Suppose John makes an outfit consisting of a shirt, a pair of pants, and a pair of shoes. How many different outfits does he have?

Copyright © Areteem Institute. All rights reserved.

(b) Repeat (a) if John *can* wear shorts instead of pants.

(c) Now suppose John can wear shorts or pants as in (b), *but* if he wears shorts, he will also wear a hat and possibly a jacket.

Problem 6.2 Suppose you have a group of 6 people. How many different photographs are there of everyone lined up if:

(a) all the people look different?

(b) 2 of the people are identical twins who have dressed identically?

(c) 2 of the people are a couple and must stand next to each other?

(d) 2 of the people are sworn enemies and cannot stand next to each other?

Problem 6.3 There are 30 lottery balls labeled from 1 to 30.

(a) How many ways are there to draw 5 lottery balls, in order one after another, if we do not replace the ball after each pick? (That is, it is not possible to pick the same ball more than once.)

(b) How many ways are there to draw 5 lottery balls all at once? (That is, it is not possible to draw the same ball twice, and the 5 balls are in no particular order.)

Copyright © Areteem Institute. All rights reserved.

Problem 6.4 How many rearrangements can be made of the letters in the word BA-NANAS?

Problem 6.5 10 friends decide to play five versus five basketball, so they need to divide themselves into two teams. If there is no order associated with how the people are picked and no order associated with the teams, how many different ways can they divide themselves into the two teams?

Problem 6.6 10 points are marked on the plane. How many different triangles can be formed using these points as vertices if no three of the points are in a straight line?

Problem 6.7 Suppose you write out the numbers $1 - 1000$: $1, 2, 3, 4, \ldots, 1000$.

(a) How many digits have you written in total?

(b) What is the sum of all the numbers written?

(c) What is the sum of all the digits written?

Problem 6.8 Suppose a pizza place has 5 toppings available. You want to order 2 different 3-topping pizzas. Suppose repeated toppings are not allowed on a single pizza, and the order of the toppings on a pizza does not matter. If you only care which two pizzas you get, how many ways are there to make the order?

Copyright © ARETEEM INSTITUTE. All rights reserved.

Problem 6.9 It is time for Dennis to make a new password. He's not too creative so he decides to create a password with his name and birthday, which is May 25. He wants to use the letters of his name (in order) and the letters/symbols of his birthday (in order). That is, the password could start with either D or M; if it starts with D the next letter could be M or e, and if it starts with M the next letter could be D or a. Examples of possible passwords are MayDennis25 or DenMaynis25 or DeManyni2s5. A password of 25DennisMay or nisMay25Den is NOT allowed. How many possible passwords could Dennis make?

Problem 6.10 Carrie invites 9 of her friends for dinner. Carrie will sit with four of the friends at the first circular table, while the other 5 friends sit around the second table. If all the seats at both tables are indistinguishable, how many seating arrangements are there?

Copyright © ARETEEM INSTITUTE. All rights reserved.

6.2   Quick Response Questions

Problem 6.11 Suppose a 8×8 checkerboard is colored red and black. How many ways are there to put a red checker and a black checker on the board so that the black checker is in a red square and the red checker is in a black square?

Problem 6.12 Suppose a 8×8 checkerboard is colored red and black. How many ways are there to put a red checker and a black checker on the board so that the checkers are in two squares of different colors?

Problem 6.13 Calculate $\dfrac{4!}{9!} \times \dfrac{12!}{6!}$.

Problem 6.14 Which of the following expressions is equal to $8 \times 7 \times 6$?

(A) $8!$

(B) $8! - 5!$

(C) $\dfrac{8!}{5!}$

(D) $8! \cdot 5!$

Problem 6.15 Calculate $\dbinom{7}{3}$

Copyright © ARETEEM INSTITUTE. All rights reserved.

Problem 6.16 Which of the following expressions is equal to $\binom{n}{k}$?

(A) $\binom{k}{n}$

(B) $\binom{n}{n-k}$

(C) $\binom{n+k}{n-k}$

(D) None of the above

Problem 6.17 Which of the following expressions is equal to $\binom{n}{2}$?

(A) n

(B) 1

(C) $\frac{n(n-1)}{2}$

(D) $\frac{n!}{(n-2)!}$

Problem 6.18 How many words are there that contain 4 A's and 5 B's?

Problem 6.19 Suppose you have 3 males and 3 females. How many ways are there to line them all up so that the males are all together and the females are all together?

Problem 6.20 Suppose we want to write the letters A, B, C, D along the outside of a frisbee (a circular disk). How many ways are there to do so? List all the outcomes.

Copyright © ARETEEM INSTITUTE. All rights reserved.

6.3 Practice Questions

Problem 6.21 A restaurant has 5 choices of appetizers, 3 choices for desserts, and for entree they offer 7 types of burgers and 4 types of salads. How many different meals (one appetizer, one entree, one dessert) are available?

Problem 6.22 Suppose the Martian alphabet has 10 letters, and a Martian word is any sequence of 4 letters, allowing repeats.

(a) How many Martian words are there in total?

(b) How many Martian words are there with no repeating letters?

(c) How many Martian words are there with at least one pair of repeating letters?

Problem 6.23 Suppose 5 people say $A, B, C, \ldots, H$ run a race.

(a) How many different outcomes are there for the race? (Suppose there are no ties!)

(b) How many different ways are there to give out Gold, Silver, and Bronze Medals?

(c) Suppose the top 3 finishers in the race advance to the next race. How many different groups can advance?

Copyright © ARETEEM INSTITUTE. All rights reserved.

Problem 6.24 How many rearrangements can be made of the letters in *MISSISSIPPI*?

Problem 6.25 Suppose you have 3 men and 3 women at a dance class. We want to divide them into 3 pairs, where the pairs are in no particular order. How many ways to do so are there if

(a) each pair is a male and a female?

(b) there are no restrictions in how the pairs are chosen?

Problem 6.26 10 points are marked on the plane. How many different triangles can be formed using these points as vertices if 5 points are on one line, and the other 5 points are on another line parallel to the first?

Problem 6.27 Suppose that Billy is reading a book. If you wrote out all the page numbers that Billy has read $(1, 2, 3, \ldots)$ you would end up writing 648 total digits. What page number is Billy on?

Problem 6.28 A pizza place has 5 toppings available. You want to order 3 different 2-topping pizzas. Order of the toppings on a pizza does not matter. If toppings are allowed to be repeated and you only care which 3 pizzas you get, how many ways to make the order are there?

Copyright © ARETEEM INSTITUTE. All rights reserved.

Problem 6.29 Dennis used his name and birthday to make his computer password. He ended up choosing the password MayDen25nis. It is now time to change his password. He decided to change his password by switching the case of some of the letters (lowercase to uppercase or vice versa), but not changing the order of anything. For example, MAYDEN25nis or mAYdEN25NIS are both possible. How many different passwords could Dennis create? Remember that Dennis is changing his password.

Problem 6.30 Pam and Jim go out to dinner with 5 of their friends. They sit at a circular table. How many seating arrangements are there if Pam and Jim sit next to each other an we only care about how the guests are arranged amongst themselves? (None of the seats are special.)

Copyright © ARETEEM INSTITUTE. All rights reserved.

7. Sets and Functions

Sets

- A *set* is an unordered collection of objects, without repetitions. We call the members of a set its *elements*.
- For example, $\{1,4,5\}$ is a set of 3 numbers. We have $\{1,4,5\} = \{4,1,5\} = \{1,1,4,5\}$.
- A set B is a *subset* of A, written $B \subseteq A$ if every element of B is an element of A.
- The *empty set* is the set with no elements. It is denoted by $\emptyset$ or sometimes $\{\}$.
- If a set A is finite, we use the notation $n(A)$ to denote the number of elements in A.
- Note: Sets can contain more than just numbers. For example, we could have the set of all states in the US, or the set of all words starting with the letter A, etc.

Combining Sets

- In counting and probability, the set of all possible outcomes of an experiment the *sample space*, and denote it by Ω (the capital Greek letter omega).
- We will call a subset A of Ω (written $A \subseteq \Omega$) an *event*.
- The complement of an event A is the collection of all elements of Ω not in A, denoted A^c.
- $A \cup B$ denotes the elements in either A or B (or both). This is called the *union* of A and B.
- $A \cap B$ denotes the elements in both A and B. This is called the *intersection* of A and B.

Principle of Inclusion-Exclusion

Copyright © ARETEEM INSTITUTE. All rights reserved.

- The *Principle of Inclusion-Exclusion* (or PIE for short) helps calculate the size of the union of two or more sets.
- For two sets, we have $n(A \cup B) = n(A) + n(B) - n(A \cap B)$.
- Venn Diagrams are useful in remembering and visualizing the Principle of Inclusion-Exclusion.

Functions and Injections

- Recall that a function is just a mapping between two sets A and B.
- Functions are often given by equations, but this is not necessary. For example, "the first letter of your first name" can be thought of as a function from the set of all names to the set $\{A, B, C, \ldots, Z\}$ of letters of the alphabet.
- A function is *injective* or *one-to-one* if every output has exactly one input. For example, $f(x) = x$ is injective, but $f(x) = x^2$ is not.

Surjections and Bijections

- A function is *surjective* or *onto* if every value in the target set is the output of some input. That is, the range of the function is the same as the target set. As functions from the real numbers to the real numbers $f(x) = x$ is surjective, but $f(x) = x^2$ is not surjective.
- If a function is injective and surjective it is called a *bijection*.
- In other words, given two sets A, B, a bijection from A to B is a one-to-one correspondence between members of A and members of B. That is, every element in a "matches" up with exactly one element in "B".

7.1 Example Questions

Problem 7.1 You roll 2 four-sided dice. Let A be the event that the first die is a 4, and B the event that the sum of the two rolls is 6.

(a) Write out the sample space Ω representing a list of all possible outcomes of rolling 2 four-sided dice. What is $n(\Omega)$?

Copyright © ARETEEM INSTITUTE. All rights reserved.

(b) List the outcomes in event A. What is $n(A)$?

(c) List the outcomes in event B. What is $n(B)$?

(d) List the outcomes in event $A \cap B$. What is $n(A \cap B)$?

Problem 7.2 How many numbers less than 1000 are divisible by 11 but not by 9?

Problem 7.3 Students in Areteem Institute were asked which pets (dogs or cats) do they have. In a survey of 50 students, 10 of them answered "No pets", 30 answered "a dog" and 20 answered "a cat". How many students have both a cat and dog?

Problem 7.4

(a) Work out and write out the PIE formula for 3 sets A, B, C.

(b) How many terms will the PIE formula for 4 sets A, B, C, D have?

Problem 7.5 How many positive integers ≤ 1000 are a perfect square, cube, fourth, or fifth power?

Problem 7.6 The following functions are all between the real numbers and itself ($\mathbb{R} \to \mathbb{R}$).
For each of the functions: (i) is it injective/one-to-one?, (ii) what is its range?

Copyright © Areteem Institute. All rights reserved.

(a) $y = x^3 + 3$

(b) $y = x^3 - x$

(c) $y = 2^x$

Problem 7.7 Let $A = \{1, 2, 3, 4\}$ and $B = \{1, 2, 3\}$. Answer each of the following. Explain how you have already answered this type of question in the past.

(a) How many total functions are there from A to B? from B to A?

(b) How many bijections are there from A to B?

(c) How many injections are there from B to A?

Problem 7.8 Let $A = \{1, 2, 3, 4\}$ and $B = \{1, 2, 3\}$. How many surjections are there from A to B?

Problem 7.9 Suppose you have a set $S = \{1, 2, 3, \ldots, 20\}$. You want to choose A, B, C such that $A \cup B \cup C = S$ and $A \cap B \cap C = \emptyset$. (Remember $\emptyset = \{\}$ is the empty set, which contains no elements.) How many ways an this be done

(a) if we also assume $A \cap B = A \cap C = B \cap C = \emptyset$? (Under these conditions, A, B, C *partitions S.*)

Copyright © ARETEEM INSTITUTE. All rights reserved.

(b) in total?

Problem 7.10 Let $S = \{1, 2, 3, 4, 5\}$. How many 5-digit numbers can be formed from members of S with no repeated digits and the 2 next to 1 or 3?

Copyright © ARETEEM INSTITUTE. All rights reserved.

7.2   Quick Response Questions

Problem 7.11 Let $A = \{1,2,3,4,5\}$, $B = \{4,5,6,7,8\}$. Which of the following is $A \cap B$?

(A) $\{1,2,3,4,5\}$
(B) $\{4,5,6,7,8\}$
(C) $\{1,2,3,4,5,6,7,8\}$
(D) $\{4,5\}$

Problem 7.12 Let $A = \{1,2,3,4,5\}$, $B = \{4,5,6,7,8\}$. What is $A \cup B$?

(A) $\{1,2,3,4,5\}$
(B) $\{4,5,6,7,8\}$
(C) $\{1,2,3,4,5,6,7,8\}$
(D) $\{4,5\}$

Problem 7.13 Let $A = \{1,2,3,4,5\}$, $B = \{4,5,6,7,8\}$. What is $(A \cup B) \cap B$?

(A) $\{1,2,3,4,5\}$
(B) $\{4,5,6,7,8\}$
(C) $\{1,2,3,4,5,6,7,8\}$
(D) $\{4,5\}$

Problem 7.14 In a class of 45 students, 26 like to play cricket and 21 like to play football. Also, each student likes to play at least one of the two games. How many students like to play both cricket and football ?

Problem 7.15 The Venn Diagram for two sets A, B has 3 sections. How many sections does a Venn Diagram for three sets A, B, C have?

Copyright © ARETEEM INSTITUTE. All rights reserved.

Problem 7.16 Which of the following is true for the function $y = x - x^5$?

(A) It is injective and surjective.
(B) It is injective but not surjective.
(C) It is surjective but not injective.
(D) It is neither injective nor surjective.

Problem 7.17 Which of the following is true for the function $y = 2 - \left(\dfrac{1}{2}\right)^x$?

(A) It is injective and surjective.
(B) It is injective but not surjective.
(C) It is surjective but not injective.
(D) It is neither injective nor surjective.

Problem 7.18 Which of the following is true for the function $y = |x - 2|$?

(A) It is injective and surjective.
(B) It is injective but not surjective.
(C) It is surjective but not injective.
(D) It is neither injective nor surjective.

Problem 7.19 Which of the following is true for the function $y = 2x + 3$?

(A) It is injective and surjective.
(B) It is injective but not surjective.
(C) It is surjective but not injective.
(D) It is neither injective nor surjective.

Problem 7.20 Let A be the set of all primes and B be the set of all even numbers. What is the result when you add up all the elements in $A \cap B$?

Copyright © ARETEEM INSTITUTE. All rights reserved.

7.3 Practice Questions

Problem 7.21 Let A be the set of perfect squares in the set of whole numbers from 1 to 100 (inclusive) and let B be the set of perfect cubes in the set of whole numbers from 1 to 100 (inclusive).

(a) List the elements in A. List the elements in B. What are $n(A)$ and $n(B)$?

(b) List the elements in $A \cap B$. List the elements in $A^c \cap B$. What are $n(A \cap B)$ and $n(A^c \cap B)$?

Problem 7.22 Consider the set of whole numbers from 1 to 100 (inclusive). How many are perfect square or perfect cubes but not both a perfect square and a perfect cube?

Problem 7.23 In a school there are 30 teachers who teach mathematics or physics. Of these, 12 teach mathematics and 6 teach both physics and mathematics. How many teach physics?

Problem 7.24 Write out the PIE formula for four sets A, B, C, D.

Problem 7.25 How many numbers between 1 and 100 (inclusive) are a multiple of either 5, 7, 11, or 13?

Problem 7.26 Any line of the form $y = mx + b$ or quadratic of the form $y = ax^2 + bx + c$ can be thought of as a function from the real numbers to the real numbers ($\mathbb{R} \to \mathbb{R}$).

Copyright © ARETEEM INSTITUTE. All rights reserved.

(a) Explain using the graph why any line with equation $y = mx + b$ and slope $m \neq 0$ is both injective and surjective.

(b) Explain using the graph why any quadratic with equation $y = ax^2 + bx + c$ (and $a \neq 0$) is neither injective nor surjective.

Problem 7.27 Let $A = \{1, 2\}$ and $B = \{1, 2, 3, 4, 5\}$. Answer each of the following. Explain how you have already answered this type of question in the past.

(a) How many total functions are there from A to B? from B to A?

(b) How many injections are there from A to B?

Problem 7.28 Let $A = \{1, 2\}$ and $B = \{1, 2, 3, 4, 5\}$. How many surjections are there from B to A? Hint: Consider complementary counting.

Problem 7.29 Suppose you have a set $S = \{1, 2, 3, \ldots, 20\}$. You want to choose A, B, C such that $A \cup B \cup C = S$ and $A \cap B \cap C = \emptyset$. How many ways are there to do so if $A \neq \emptyset$?

Problem 7.30 Suppose that students take three tests in a course and that exactly 11 students get A's on each exam. How many students must get A's on all three exams if exactly 9 students get A's on any two exams and 14 students get an A on at least one exam?

Copyright © ARETEEM INSTITUTE. All rights reserved.

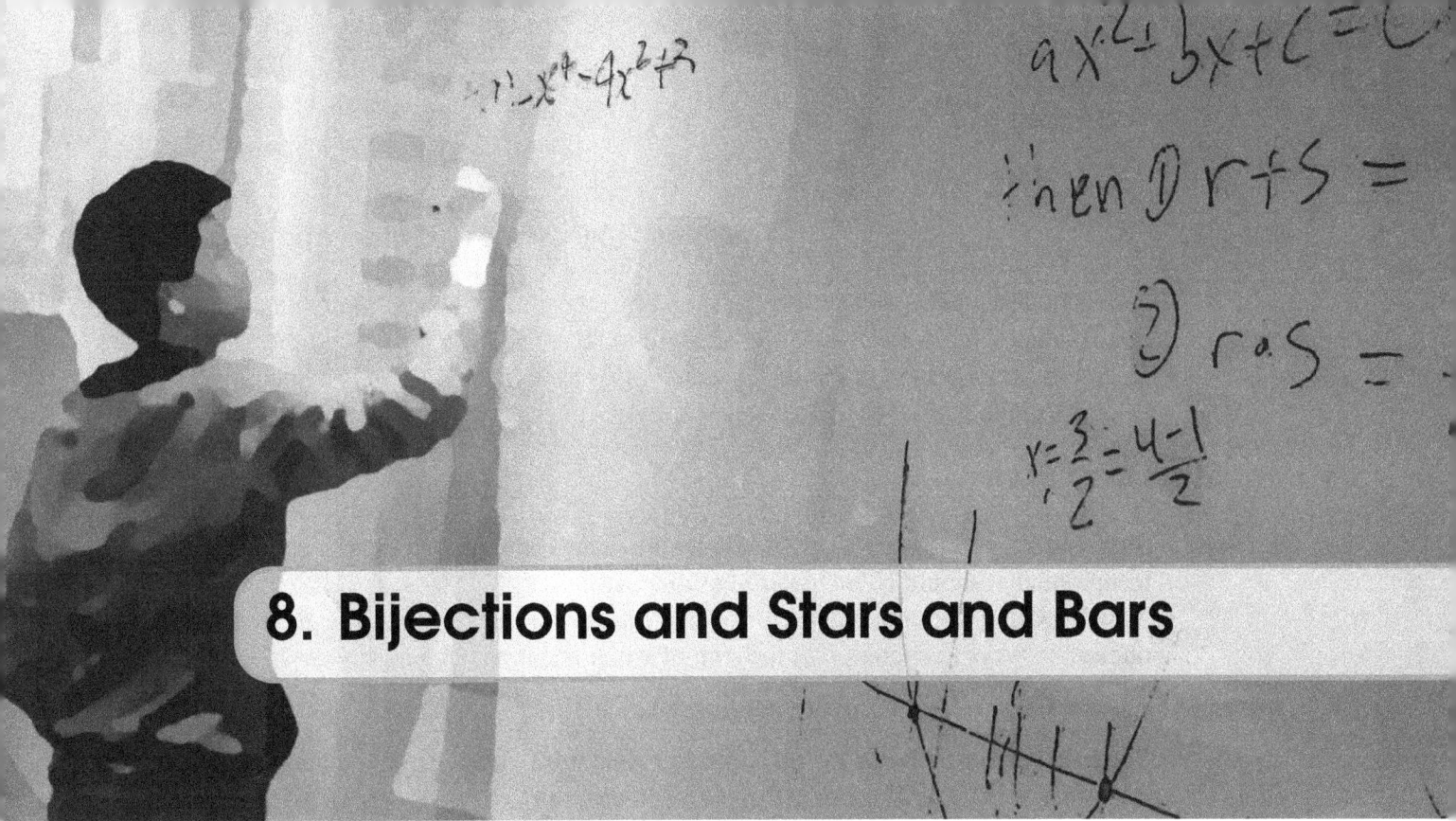

8. Bijections and Stars and Bars

Basic Review of Sets

- A *set* is an unordered collection of elements, without repetitions. Sets are often denoted A, B, C, etc.
- If A, B are sets, then
  - $n(A), n(B)$ denotes the size of A and B respectively.
  - $A \cap B$ is the *intersection* of A and B. $A \cap B$ consists of all the elements both in A and in B.
  - $A \cup B$ is the *union* of A and B. $A \cup B$ consists of all the elements in A or in B (or in both).
  - If every element of A is an element of B, we say A is a subset of B and write $A \subseteq B$.

Review of Functions

- A function is just a mapping between two sets A and B.
- A function is *injective* or *one-to-one* if every output has exactly one input. For example, $f(x) = x$ is injective, but $f(x) = x^2$ is not.
- A function is *surjective* or *onto* if every value in the target set is the output of some input. That is, the range of the function is the same as the target set. As functions from the real numbers to the real numbers $f(x) = x$ is surjective, but $f(x) = x^2$ is not surjective.
- If a function is injective and surjective it is called a *bijection*.
- In other words, given two sets A, B, a bijection from A to B is a one-to-one

---

Copyright © Areteem Institute. All rights reserved.

correspondence between members of A and members of B. That is, every element in a "matches" up with exactly one element in "B".

- Bijections are useful in counting because of this basic fact: If there is a bijection from A to B, then $n(A) = n(B)$.

Stars and Bars

- Stars and Bars (Balls and Urns, etc., there are many different names) is a counting technique that should be memorized as all costs.
- **Non-Negative Version**: Given a positive integer n, and positive integer k, the number of ways to express n as the sum of k non-negative integers ($n = a_1 + a_2 + \cdots + a_k$ where $a_1, a_2, \ldots, a_k$ are non-negative) is $\binom{n+k-1}{n}$.
- **Positive Version**: Given a positive integer n, and positive integer k, the number of ways to express n as the sum of k positive integers ($n = a_1 + a_2 + \cdots + a_k$ where $a_1, a_2, \ldots, a_k$ are positive) is $\binom{n-1}{k-1}$.

8.1 Example Questions

Problem 8.1 Suppose below is a map of a city you want to travel from A to B.

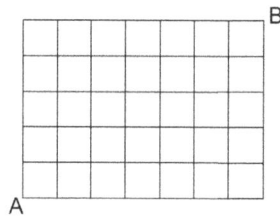

(a) If each square in the diagram is a square block what is the minimum number of blocks it takes to get from A to B?

(b) How many paths of shortest length are there from A to B?

Copyright © ARETEEM INSTITUTE. All rights reserved.

Problem 8.2 More Paths

(a) Consider the grid below.

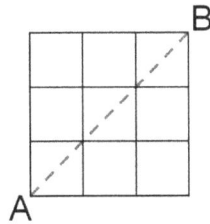

Consider paths from A to B where each step is either right (R) or up (U). How many such paths do not go below the dotted line? List out these paths.

(b) Consider the set of all words made up of 3 R's and 3 U's. Describe the properties one of these words should have to correspond to one of the paths as in part (a).

Problem 8.3 Let $A = \{1, 2, 3, 4\}$ and $B = \{0, 1\}$.

(a) How many functions from A to B are there?

(b) How many subsets of A are there?

(c) Explain your answers in (a) and (b) using a bijection.

Problem 8.4 Let $A = \{1, 2, 3, 4, 5\}$.

Copyright © ARETEEM INSTITUTE. All rights reserved.

(a) Give an example of a bijection between subsets of A of size two and subsets of A of size three.

(b) How many such bijections are there in part (a)?

Problem 8.5 Let $A = \{1, 2, 3, 4\}$.

Find a bijection between subsets of A of size 2 and two numbers chosen (repetition allowed) from $\{1, 2, 3\}$.

Problem 8.6 Stars and Bars

(a) You want to arrange 6 stars ($*$'s) and 3 bars ($|$'s) in a line. How many ways are there for you to do so?

(b) How many ways are there to choose non-negative integers a, b, c, d such that $a + b + c + d = 6$? Explain how your answer in part (a) can help.

Problem 8.7 Only Positive Numbers

(a) Consider 5 plus signs in a row: $\underline{+}\,\underline{+}\,\underline{+}\,\underline{+}\,\underline{+}$. How many ways are there to choose 3 plus signs to remove? For example, if you remove the first, second, and fourth you are left with $\_\,\_\,\underline{+}\,\_\,\underline{+}$.

(b) How many ways are there to choose positive integers a, b, c such that $a + b + c = 6$? Explain how your answer in part (a) can help.

Copyright © ARETEEM INSTITUTE. All rights reserved.

Problem 8.8 How many ways are there to choose non-negative integers a, b, c, d such that $a + b + c + d = 6$ with $a = 1$ and $b \geq 2$?

Problem 8.9 Suppose 5 people get in an elevator on Floor 0. The people leave the elevator somewhere between (inclusive) Floors 1 and Floor 5.

(a) If we only care about how many people get of at each floor, how many ways can the people get off?

(b) If we only care about what collection of floors the elevator stops on, how many different collections are there?

Problem 8.10 You line up 11 cards in a row. 8 of the cards are black and identical. The other 3 cards are red and numbered 1, 2, and 3. How many different ways to line up the 11 cards are there if each of the red cards is separated by at least 2 black cards?

Copyright © ARETEEM INSTITUTE. All rights reserved.

8.2 Quick Response Questions

Problem 8.11 Consider paths from A to B where each step is either right or up in the grid below.

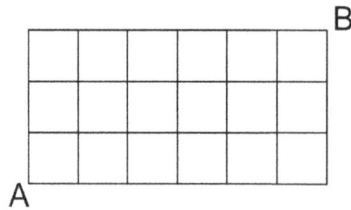

How many paths are there in total?

Problem 8.12 George flips a coin 7 times. How many outcomes are there with more heads than tails?

Problem 8.13 Which of the following represents the equation $1 + 0 + 3 + 2 = 6$ using stars and bars?

(A) $|*||***|**||$
(B) $|*|*|*|*|*|$
(C) $*||***|**|$
(D) $|**|||*||*$

Copyright © ARETEEM INSTITUTE. All rights reserved.

Problem 8.14 Suppose you roll six dice and you only care about what you rolled (not which die has which number). Which of the following could you use to represent the outcome in terms of stars and bars if you get two 1's, three 4's and a 5?

(A) $**|***||||*|||||$
(B) $**|||***|*|$
(C) $|**|||***|*||$
(D) $||*|||****|*****$

Problem 8.15 How many ways can you express 6 as the sum of three non-negative integers?

Problem 8.16 How many ways can you express 6 as the sum of three positive integers?

Problem 8.17 Alice, Bob, Charles, and Desiree are 4 students comparing the days of the week on which they were born. In total how many possibilities are there for the day of the week each is born?

Problem 8.18 Alice, Bob, Charles, and Desiree are 4 students comparing the days of the week on which they were born. How many possible outcomes are there if we don't care about who was born on a given day? That is, we only care how many of them are both on each day of the week.

Problem 8.19 Alice, Bob, Charles, and Desiree are 4 students comparing the days of the week on which they were born. How many possible outcomes are there if they were all born on different days? Assume we don't care about who was born on on a given day.

Copyright © ARETEEM INSTITUTE. All rights reserved.

Problem 8.20 Alice, Bob, Charles, and Desiree are 4 students comparing the days of the week on which they were born. Suppose we do care about who was born on a given day. How many possible outcomes are there if they were all born on different days?

Copyright © ARETEEM INSTITUTE. All rights reserved.

8.3 Practice Questions

Problem 8.21 Below is a map of city blocks and you travel from A to B moving only up or right on the grid.

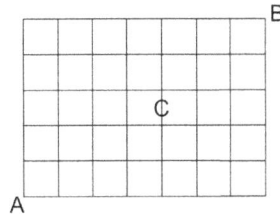

How many of these paths travel through the block marked with a C?

Problem 8.22 In an earlier problem we looked at paths from A to B (each step is R or U) that did not go below the dotted line in the diagram below.

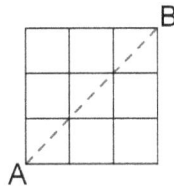

Consider now filling in the numbers $1, 2, 3, 4, 5, 6$ in the 2×3 table below:

so that each row (left to right) is increasing and each column (top to bottom) is also increasing. How many ways are there to fill in these 6 numbers? Compare this to the number of paths. Do you notice any pattern?

Copyright © ARETEEM INSTITUTE. All rights reserved.

Problem 8.23 Let $A = \{1,2,3,4\}$ and $B = \{0,1\}$. There are $2^4 = 16$ functions from A to B. There are also $2^4 = 16$ sequences of length 4 made up of elements from B. Explain this by finding a bijection from the 16 functions to the 16 sequences. That is, explain how the functions and sequences match up.

Problem 8.24 Let $A = \{1,2,3,4,5\}$

(a) How many subsets of A are there?

(b) Verify that $\binom{5}{0} + \binom{5}{1} + \binom{5}{2} + \binom{5}{3} + \binom{5}{4} + \binom{5}{5} = 32$.

(c) Explain using subsets and cases why $\binom{5}{0} + \binom{5}{1} + \binom{5}{2} + \binom{5}{3} + \binom{5}{4} + \binom{5}{5}$ represents the number of subsets of A.

Problem 8.25 Find a bijection between subsets of $\{1,2,3,4,5\}$ of size 3 and 3 numbers chosen (repetition allowed) from $\{1,2,3\}$.

Problem 8.26 Consider non-negative integers a,b,c,d,e such that $a+b+c+d+e = 10$.

(a) If we want to express this equation using stars and bars, how many stars do we need? how many bars?

(b) How many different ways are there to choose the 5 integers a,b,c,d,e?

Copyright © ARETEEM INSTITUTE. All rights reserved.

Problem 8.27 How many ways are there to write 6 as the sum of one or more positive integers?

Problem 8.28 Suppose you have 30 identical balls and 6 numbered boxes. How many ways are there to put the balls into the boxes if:

(a) there are no restrictions?

(b) each box has at least two balls?

(c) the first box has exactly 10 balls?

Problem 8.29 Beagel likes bagels, and he went to the Bagel Shop to buy 6 bagels for breakfast. The Bagel Shop sells 3 types of bagels: sourdough, blueberry, and sesame seeds. How many ways are there to buy the bagels so that Beagel gets at least 2 types?

Problem 8.30 For their birthday party, identical twins Harry and Kerry dress up in identical outfits so they can confuse their friends by looking the same. They line up for a photo with their 5 best friends at the end of the party. They do not stand on the ends of the line, nor do they stand next to each other. How many different photos are possible?

Copyright © Areteem Institute. All rights reserved.

9. Probability

Basic Review of Sets

- A *set* is an unordered collection of elements, without repetitions. Sets are often denoted A, B, C, etc.
- If A, B are sets, then
    - $n(A), n(B)$ denotes the size of A and B respectively.
    - $A \cap B$ is the *intersection* of A and B. $A \cap B$ consists of all the elements both in A and in B.
    - $A \cup B$ is the *union* of A and B. $A \cup B$ consists of all the elements in A or in B (or in both).
    - If every element of A is an element of B, we say A is a subset of B and write $A \subseteq B$.

Probability (Classical Model)

- Suppose Ω is a finite sample space and every outcome in Ω is equally likely. If $A \subseteq \Omega$, then

$$\text{the probability of } A = P(A) = \frac{\text{want}}{\text{total}} = \frac{n(A)}{n(\Omega)}.$$

- For example, if we flip a fair coin twice, then $\{HH, HT, TH, TT\}$ is a sample space where every outcome is equally likely. However, the sample space $\{\text{two heads, one head and one tail, two tails}\}$ is a sample space but each outcome is *not* equally likely.

Axioms (Rules) of Probability

Copyright © ARETEEM INSTITUTE. All rights reserved.

- Suppose that Ω is a sample space and $A, B \subseteq \Omega$ are events.
  Pr1. $P(A) \geq 0$.
  Pr2. $P(\Omega) = 1$.
  Pr3. If A and B are *disjoint* (that is, $A \cap B = \emptyset$), then $P(A \cup B) = P(A) + P(B)$.
- Additional properties which can be proven from the axioms.
  Pr4. $P(\emptyset) = 0$.
  Pr5. $0 \leq P(A) \leq 1$.
  Pr6. $P(A^c) = 1 - P(A)$.
  Pr7. $P(A \cup B) = P(A) + P(B) - P(A \cap B)$.

9.1 Example Questions

Problem 9.1 You roll 2 four-sided dice. Let A be the event that the first die is a 4, and B the event that the sum of the two rolls is 6.

(a) What is $P(A)$?

(b) What is $P(B)$?

(c) What is $P(A \cap B)$?

Problem 9.2 Suppose you flip a fair coin 6 times.

(a) Find the probability of exactly 4 heads.

(b) Find the probability of at least one tails.

Copyright © Areteem Institute. All rights reserved.

Problem 9.3 A dealer starts with only the 4 aces (one of each suit) from a deck of cards. They deal you 2 of the cards. Let A be the event that you get one heart and one diamond. Let B be the event that you get a spade.

(a) Assume the cards are dealt to you in order (that is, a first card and a second card). Find $P(A)$ and $P(B)$. Hint: You may want to write out a sample space.

(b) Assume the cards are not dealt in order (that is, you are dealt two cards at once). Find $P(A)$ and $P(B)$. Hint: You may want to write out a sample space.

(c) Compare your answers in parts (a) and (b). Can you explain the outcome?

Problem 9.4 You have 5 red, numbered 1 through 5, and 8 green balls, numbered 1 through 8. You pick 5 without replacing the balls. For each of the events below, find the probability. Note: It is best to think of all 5 balls being picked at once.

(a) What is the probability you get 3 green and 2 red balls?

(b) What is the probability that all the balls you pick are the same color?

Problem 9.5 You have 3 red balls, numbered 1 through 3, and 4 green balls, numbered 1 through 4. You pick 5 balls, one by one, replacing the ball after each pick. (Thus you can pick the same ball more than once.)

(a) How many outcomes are there where you pick 2 red and 3 green balls?

Copyright © ARETEEM INSTITUTE. All rights reserved.

(b) What is the probability of picking 2 red and 3 green balls?

Problem 9.6 Assuming only the axioms Pr1 - Pr3:
Pr1. $P(A) \geq 0$.
Pr2. $P(\Omega) = 1$.
Pr3. If A and B are *disjoint* (that is, $A \cap B = \emptyset$), then $P(A \cup B) = P(A) + P(B)$.
Prove:

(a) Pr6: $P(A^c) = 1 - P(A)$.

(b) Pr4: $P(\emptyset) = 0$.

Problem 9.7 Emily builds an unfair 6-sided die. The probability of rolling a 1 is the same as rolling a 3. Rolling a 4 is twice as likely as rolling a 1 and rolling a 5 is 10% more likely than rolling a 1. If the probability of rolling a 2 is 0.2 and rolling a 6 is 0.1, find the probability of rolling each of the six sides of the die.

Problem 9.8 Probability Venn Diagram Practice

(a) If $P(A) = 0.5$, $P(B) = 0.7$, what are the maximum and minimum possible values of $P(A \cap B)$?

(b) Suppose $P(A) = 0.5$, $P(B) = 0.7$, $P(A^c \cap B^c) = 0.2$. Find $P(A \cap B^c)$ and $P(B \cap A^c)$.

Problem 9.9 Suppose you flip a fair coin 6 times.

Copyright © ARETEEM INSTITUTE. All rights reserved.

(a) Find the probability of no two heads in a row and no two tails in a row.

(b) Find the probability you get more heads than tails.

Problem 9.10 Suppose you roll a fair die 4 times. What is the probability the sum of all 4 rolls is 10?

Copyright © ARETEEM INSTITUTE. All rights reserved.

9.2   Quick Response Questions

Problem 9.11 Suppose you randomly pick an integer between 1 and 100 (inclusive). The probability that your number is a multiple of 3 is $\dfrac{P}{Q}$ in lowest terms. What is $Q - P$?

Problem 9.12 Suppose you randomly pick an integer between 1 and 100 (inclusive). The probability that your number is a multiple of 3 or 5 is $\dfrac{P}{Q}$ in lowest terms. What is $Q - P$?

Problem 9.13 A jar has 1 red, 2 green, and 3 yellow balls. Alice and Bob each pick one ball at random (at the same time so they cannot both get the same ball). The probability they both get yellow balls is $\dfrac{P}{Q}$ in lowest terms. What is $Q - P$?

Problem 9.14 A jar has 1 red, 2 green, and 3 yellow balls. Alice and Bob each pick one ball at random (at the same time so they cannot both get the same ball). The probability they get balls of different colors is $\dfrac{P}{Q}$ in lowest terms. What is $Q - P$?

Problem 9.15 Suppose $\Omega = \{1, 2, 3, 4\}$ and $P(1) = 0.1, P(3) = 0.4$ and $P(2) = P(4)$. Find $P(4)$.

Problem 9.16 Suppose you have an unfair 6-sided die with a 50% chance of rolling a one. If the other 5 sides are all equally likely to occur, the probability that you get an odd number when you roll the die once is $\dfrac{P}{Q}$ in lowest terms. What is $Q - P$?

Copyright © Areteem Institute. All rights reserved.

Problem 9.17 Suppose $P(A) = 0.6$, $P(B) = 0.5$ and $P(A^c \cap B^c) = 0.3$. Find $P(A \cap B)$.

Problem 9.18 You roll a fair six-sided die twice. The probability the sum is 8 can be written as $\dfrac{P}{36}$. What is P?

Problem 9.19 You roll a fair six-sided die three times. The probability the sum is 8 can be written as $\dfrac{P}{216}$. What is P?

Problem 9.20 You roll a fair six-sided die four times. The probability that each roll is at least 3 and the sum is 12 can be written as $\dfrac{P}{1296}$. What is P?

Copyright © ARETEEM INSTITUTE. All rights reserved.

9.3 Practice Questions

Problem 9.21 You roll 2 six-sided dice. Let A be the event that the first die is a 4, and B the event that the sum of the two rolls is 6.

(a) What is $P(A)$?

(b) What is $P(B)$?

Problem 9.22 Suppose you flip a coin 10 times. What is the probability you get ≥ 3 heads?

Problem 9.23 Suppose you have two sodas (each chosen from Coke, Sprite, Fanta). Let A be the event you have two of the same soda.

(a) Assume the sodas are bought in order (that is, a first soda and a second soda). Find $P(A)$. (It will be helpful to write out a sample space!)

(b) Assume the sodas are not bought in order (that is, you buy them both at once). Find $P(A)$. (It will be helpful to write out a sample space!)

(c) Compare your answers in parts (a) and (b). Can you explain the outcome?

Problem 9.24 You have 5 red balls, numbered 1 through 5, 3 green balls, numbered 1 through 3, and 5 yellow balls, numbered 1 through 5. You pick 5 without replacing the balls (so it is best to think of all 5 being chosen at once).

Copyright © ARETEEM INSTITUTE. All rights reserved.

(a) What is the probability you get exactly 2 red balls?

(b) What is the probability that all the balls you pick are the same color?

Problem 9.25 Bill has a 4-sided fair die with 2 sides colored gray, one side colored white, and the last side colored black. He tosses the die 5 times in a row. What is the probability that in those 5 tosses he got the color gray once and each of the colors white and black twice.

Problem 9.26 Assuming only the axioms Pr1 - Pr3:

Pr1: $P(A) \geq 0$.

Pr2: $P(\Omega) = 1$.

Pr3: If A and B are disjoint (that is, $A \cap B = \emptyset$), then $P(A \cup B) = P(A) + P(B)$.

Prove Pr7: $P(A \cup B) = P(A) + P(B) - P(A \cap B)$.

Problem 9.27 Miley builds an unfair 20-sided die. The die is numbered 1 through 20 so that the probability of rolling a given number is proportional to the number itself. For example, rolling a 9 is 3 times more likely than rolling a 3 and rolling a 12 is 6 times more likely than a 2. What is the probability of getting an even number if you roll Miley's die once?

Problem 9.28 Suppose A, B are events. If $P(A^c) = 0.3$, $P(B) = 0.5$, and $P(A^c \cup B) = 0.5$. Fill in a Venn diagram with each region labeled with its probability. That is, find $P(A \cap B^c), P(A \cap B), P(A^c \cap B), P(A^c \cap B^c)$.

Copyright © Areteem Institute. All rights reserved.

Problem 9.29 You flip a fair coin 5 times.

(a) What is the probability you get more heads than tails?

(b) What is the probability you get an even number of heads?

Problem 9.30 You roll a fair 6-sided die 4 times. What is the probability that the sum of the rolls is 10 and none of the rolls are 1?

Copyright © ARETEEM INSTITUTE. All rights reserved.

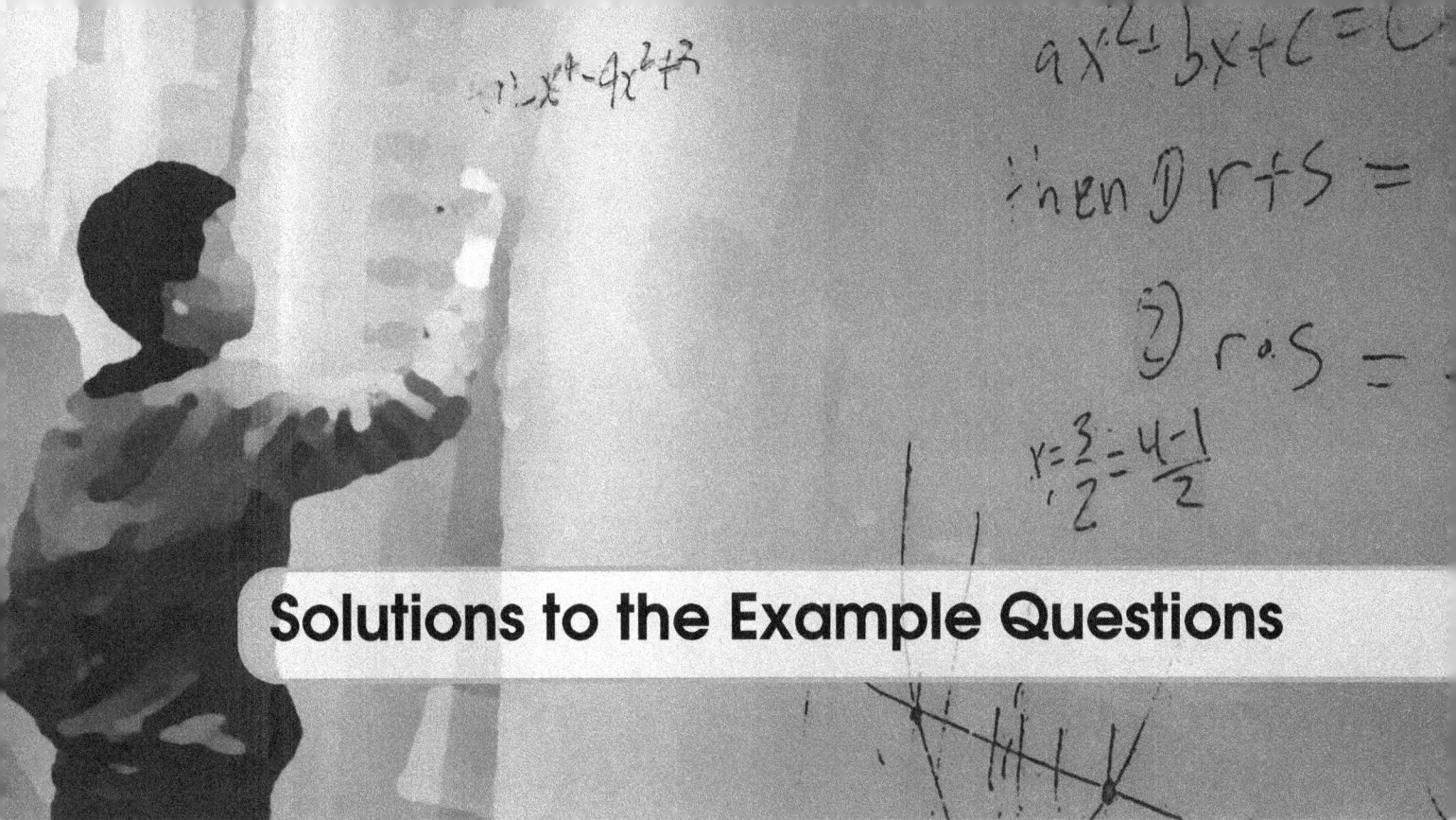

Solutions to the Example Questions

In the sections below you will find solutions to all of the Example Questions contained in this book.

Quick Response and Practice questions are meant to be used for homework, so their answers and solutions are not included. Teachers or math coaches may contact Areteem at info@areteem.org for answer keys and options for purchasing a Teachers' Edition of the course.

Copyright © ARETEEM INSTITUTE. All rights reserved.

1 Solutions to Chapter 1 Examples

Problem 1.1 Formally prove the two basic facts mentioned earlier. Note: These are not hard to prove, but it is useful to see it formally written out.

(a) If $a \mid n$ and $a \mid m$ then $a \mid n \pm m$.

Solution

By the definition of divisibility, we have $n = k \cdot a, m = j \cdot a$. Hence $n \pm m = (k \pm j) \cdot a$ so $a \mid n \pm m$.

(b) If $a \mid n$ and k is any integer, then $a \mid k \cdot n$.

Solution

By definition, $n = j \cdot a$, so $k \cdot n = (k \cdot j) \cdot a$ so $a \mid k \cdot n$.

Problem 1.2 The number 64 has the property that it is divisible by its units digit. How many whole numbers between 10 and 50 have this property?

Answer

17

Solution

They can be listed based on their last digits: 11, 21, 31, 41; 12, 22, 32, 42; 33; 24, 44; 15, 25, 35, 45; 36; 48.

Problem 1.3 Given a number $\overline{abcde}$. Prove the divisibility rule for:

(a) 5

Solution

First note that $n = \overline{abcde} = \overline{abcd} \cdot 10 + e$. 5 divides any multiple of 10 so we just need to make sure $5 \mid e$ (using our basic facts!). $5 \mid e$ if and only if $e = 0$ or 5.

(b) 9

Copyright © ARETEEM INSTITUTE. All rights reserved.

Solution

We can write $n = \overline{abcde} = 10000a + 1000b + 100c + 10d + e = 9999a + 999b + 99c + 9d + a + b + c + d + e$. Thus, $n = (1111a + 111b + 11c + d) \cdot 9 + (a + b + c + d + e)$. Hence, $9 \mid n$ if $9 \mid (a + b + c + d + e)$ as needed.

Problem 1.4 A four digit number $\overline{7a4b}$ is divisible by 18. Find the value of a and b so that this four-digit number has the largest value.

Answer

7848

Solution

Since $18 = 2 \times 9$, the number needs to be divisible by 2 and by 9. Therefore we want b to be even and $7 + a + 4 + b = 11 + a + b$ to be a multiple of 9. For the largest number we try $11 + a + b = 27$, or $a + b = 16$. Since b needs to be even, $a = 8, b = 8$ is the largest that works.

Problem 1.5 Consider the integer $\overline{2a3a1a}$.

(a) If the number is divisible by 9, what are the possible values for a?

Answer

$a = 1, 4, 7$

Solution

Using the divisibility rule for 9, the integer is divisible by 9 if $9 \mid 6 + 3a$ or $3 \mid 2 + a$. This is only true when $a = 1, 4, 7$.

(b) If the number is divisible by 11, what are the possible values for a?

Answer

$a = 2$

Copyright © Areteem Institute. All rights reserved.

Solution

Using the divisibility rule for 11, the integer is divisible by 11 if $11 \mid 6 - 3a$. This is only true when $6 - 3a = 0$ or $a = 2$.

Problem 1.6 In the multiplication problem below, A, B, C and D are distinct digits. What is $A + B$?

$$
\begin{array}{ccccc}
  &   & A & B & A & B \\
\times &   &   &   & C & D \\
\hline
  & C & D & C & D & B \\
\end{array}
$$

Answer

1

Solution

Note

$$\overline{CDCDB} = 1000\overline{CD} + 10\overline{CD} + \overline{B} = 1010\overline{CD} + \overline{B}.$$

We know $\overline{CDCDB}$ is a multiple of $\overline{CD}$, so $\overline{CD}$ must divide $\overline{B}$, thus $B = 0$. Now it is clear that $A = 1$. Therefore $A + B = 1$.

Problem 1.7 Suppose you have a 10-digit number made from four 2's, three 3's, two 4's, and one 5. Is it possible for this number to be a perfect square?

Answer

No

Solution

Regardless of how the digits are arranged, the sum of the digits is $4 \cdot 2 + 3 \cdot 3 + 2 \cdot 4 + 1 \cdot 5 = 30$. Note that 30 is a multiple of 3, so 3 is the factor of our number. However, 30 is not a multiple of 9, so 9 is not a factor of our number. Hence the number is not a perfect square.

Problem 1.8 A five-digit number $\overline{abcde}$ has digits $5, 6, 7, 8, 9$ (not necessarily in that order). Assume that $5 \mid \overline{abcde}$, $4 \mid \overline{abcd}$, $3 \mid \overline{abc}$, and $2 \mid \overline{ab}$. Find all possible values of $\overline{abcde}$.

Copyright © ARETEEM INSTITUTE. All rights reserved.

Answer

78965 or 98765

Solution

It can be determined immediately that $e = 5$. Also we know that b, d must be even digits, and so a, c are odd. So either $a = 7, c = 9$ or $c = 7, a = 9$. To make $\overline{abc}$ a multiple of 3, only $b = 8$ works. This means $d = 6$. Checking, both 78965 and 98765 will work.

Problem 1.9 A positive integer is equal to 18 times the sum of its digits. What is this number?

Answer

162

Solution

18 times the sum of a number's digits is much smaller than the number for any number ≥ 1000, so we examine numbers ≤ 999.

Let the number be $\overline{abc} = 100a + 10b + c$. We thus want $100a + 10b + c = 18(a + b + c)$ or after simplifying $82a = 8b + 17c$. Since $b, c \leq 9$,

$$82a = 8b + 17c \leq 8 \cdot 9 + 17 \cdot 9 = 225 \Rightarrow a \leq \frac{225}{82} < 3.$$

Therefore $a = 0, 1, 2$. If $a = 0$, then $b = c = 0$ which does not work. If $a = 1$ we have $82 = 8b + 17c$ or $b = \dfrac{82 - 17c}{8}$. This is only an integer if $c = 2$, and $b = \dfrac{82 - 17 \cdot 2}{8} = 6$, giving the number 162. If $a = 2$ we have $164 = 8b + 17c$ or $b = \dfrac{164 - 17c}{8}$ which is never an integer.

Hence 162 is the only such number.

Problem 1.10 Show that $\sqrt{3}$ is an irrational number.

Solution

Pretend that $\sqrt{3}$ is a rational number, so we can write $\sqrt{3} = \dfrac{r}{s}$ for integers r and s and the fraction is fully reduced. Squaring both sides and clearing the denominators we have

Copyright © Areteem Institute. All rights reserved.

$3s^2 = r^2$. This means that r^2 is a multiple of 3, so in fact $r = 3m$ for another integer m. Substituting we have $3s^2 = (3m)^2$ or $s^2 = 3m^2$ after simplifying. The same argument as before thus tells us that s^2 is a multiple of 3, so $s = 3n$ for some integer n. However, this means that $\dfrac{r}{s} = \dfrac{3m}{3n} = \dfrac{m}{n}$ so in fact the fraction was NOT reduced. This is a contradiction, so $\sqrt{3}$ cannot be rational.

Copyright © ARETEEM INSTITUTE. All rights reserved.

2 Solutions to Chapter 2 Examples

Problem 2.1 The numerical values of the years are favorite numbers of many math problems. For each of the following, (i) Find the prime factorization and (ii) Find the number of factors.

(a) 2017

Answer

2017 is prime, 2 factors

Solution

2017 is prime, so it only has 2 factors (1 and itself).

(b) 2018

Answer

$2018 = 2 \cdot 1009$, 4 factors

Solution

2018 is even, so we know it is divisible by 2. As $2018 \div 2 = 1009$ which is prime, its prime factorization is $2 \cdot 1009$. Hence 2018 has $(1+1)(1+1) = 4$ factors.

(c) 2019

Answer

$2019 = 3 \cdot 673$, 4 factors

Solution

Note $2+0+1+9 = 12$ is divisible by 3, so 2019 is divisible by 3 as well. $2019 \div 3 = 673$, which is prime. The prime factorization is thus $3 \cdot 673$ and there are $(1+1)(1+1) = 4$ factors.

(d) 2020

Copyright © Areteem Institute. All rights reserved.

Answer

$2020 = 2^2 \cdot 5 \cdot 101$, 12 factors

Solution

Note $2020 = 20 \cdot 101$. As 101 is prime and $20 = 2^2 \cdot 5$ the prime factorization is $2^2 \cdot 5 \cdot 101$. Hence 2020 has $(2+1)(1+1)(1+1) = 12$ factors.

Problem 2.2 Prove that a number has an odd number of factors if and only if it is a square.

Solution 1

Note all the factors come in pairs. For any number n, if a is a factor, so is n/a. Pairing up the factors in this way, we see that a number has an even number of factors unless $a = \sqrt{n}$ is a factor (which is then paired up with itself).

Solution 2

If $n = p_1^{e_1} p_2^{e_2} \cdots p_k^{e_k}$, we know n has $(e_1 + 1) \cdots (e_k + 1)$ factors. Note if one of these terms (the $(e_i + 1)$'s) is even, then n has an even number of factors. Else all the terms are odd, so e_i is even for all i. Hence n is a perfect square.

Problem 2.3 Find the smallest positive integer x such that

(a) $252 \cdot x$ is a perfect square.

Answer

7

Solution

Note $252 = 2^2 \cdot 3^2 \cdot 7$. To make a perfect square we only need another 7.

(b) $252 \cdot x$ is a perfect cube.

Answer

294

Copyright © ARETEEM INSTITUTE. All rights reserved.

Solution

As above $252 = 2^2 \cdot 3^2 \cdot 7$. To make a perfect cube we need $2 \cdot 3 \cdot 7^2 = 294$.

Problem 2.4 Write $10! = A \cdot B \cdot C \cdot D$ for positive integers $A \leq B \leq C \leq D$ with A a factor of B, C, and D. What is the smallest possible value of $D - A$?

Answer

72

Solution

The prime factorization of $10!$ is $2^8 \cdot 3^4 \cdot 5^2 \cdot 7$. Hence $\sqrt[4]{10!} = 2^2 \cdot 3 \cdot \sqrt[4]{5^2 \cdot 7}$. So we set $A = 2^2 \cdot 3$. If we want $D - A$ to be minimized we set

$$A = 2^2 \cdot 3 = 12, B = C = 2^2 \cdot 3 \cdot 5 = 60, D = 2^2 \cdot 3 \cdot 7 = 84.$$

Therefore $D - A = 84 - 12 = 72$.

Problem 2.5 A natural number is a multiple of 72, and has a total of 15 factors. Find the largest such number.

Answer

144

Solution

The number is a multiple of $72 = 2^3 \times 3^2$. We have that $15 = 15 \times 1 = 5 \times 3$ so since the number we want has 15 factors, it is either in the form p^{14} or $p^4 \times q^2$ for primes p, q. The number must have two prime factors, so it must be of the form $p^4 \times q^2$. Further, it must have a factor of 2^3, so we must have $p = 2, q = 3$ so the number is $2^4 \times 3^2 = 144$. Since this is the only possibility, it is the largest, so 144 is the answer.

Problem 2.6 Consider numbers n with the property that the factors of n multiply out to n^3. For example, the factors of 12 are $1, 2, 3, 4, 6, 12$ and

$$1 \cdot 2 \cdot 3 \cdot 4 \cdot 6 \cdot 12 = 1728 = 12^3.$$

In fact, 12 is the smallest such number > 1. What is the next smallest number with this property?

Copyright © ARETEEM INSTITUTE. All rights reserved.

Answer

18

Solution

Note that 12 has 6 factors, which can be paired so that

$$(1 \cdot 12) \cdot (2 \cdot 6) \cdot (3 \cdot 4) = 12^3.$$

With this reasoning, we see that any number with exactly 6 factors will have the property we want. Then note that 18 is the next number with exactly 6 factors $(1,2,3,6,9,18)$, so the answer is 18.

Problem 2.7 Find the greatest common divisor and the least common multiple of

(a) 123 and 1681.

Answer

GCD: 41, LCM: 5043.

Solution

We have $123 = 3 \cdot 41$ and $1681 = 41^2$. Therefore, $\gcd(123, 1681) = 3^0 \cdot 41^1 = 41$ and $\operatorname{lcm}(123, 1681) = 3^1 \cdot 41^2 = 5043$.

(b) 8! and 4^3.

Answer

GCD: 64, LCM: $8! = 40320$.

Solution

We have $8! = 2^7 \cdot 3^3 \cdot 5 \cdot 7$ and $4^3 = 2^6$. Therefore, $\gcd(8!, 4^3) = 2^6 \cdot 3^0 \cdot 5^0 \cdot 7^0 = 64$ and $\operatorname{lcm}(8!, 4^3) = 2^7 \cdot 3^1 \cdot 5^1 \cdot 7^1 = 8! = 40320$.

(c) $3! + 5!$ and $5! + 6!$.

Answer

GCD: 42, LCM: 2520.

Copyright © Areteem Institute. All rights reserved.

Solution

We have $3!+5! = 3!(1+5\cdot4) = 6\cdot21 = 2\cdot3^2\cdot7$ and $5!+6! = 5!(1+6) = 120\cdot7 = 2^3\cdot3\cdot5\cdot7$. Therefore, $\gcd(3!+5!,5!+6!) = 2^1\cdot3^1\cdot5^0\cdot7^1 = 42$ and $\text{lcm}(3!+5!,5!+6!) = 2^3\cdot3^2\cdot5^1\cdot7^1 = 2520$.

Problem 2.8 Consider numbers that leave a remainder of 2 when divided by 3, 4, 5, and 6.

(a) Find the smallest such number.

Answer

62

Solution

If the number is N, then $N-2$ must be a common multiple of 3, 4, 5, and 6. $\text{lcm}(3,4,5,6) = 60$, so the smallest such number is 62.

(b) Find the largest such three-digit number.

Answer

962

Solution

If the number is N, then $N-2$ must be a common multiple of 3, 4, 5, and 6. $\text{lcm}(3,4,5,6) = 60$, and in 3-digits, the largest multiple of 60 is 960. Therefore the number we want is 962.

Problem 2.9 Suppose A,B,C are integers ≥ 2 with (i) $\gcd(A,B) = 12$, (ii) $\text{lcm}(A,B) = 396$, and (iii) $\gcd(B,C) = 33$. Calculate $\gcd(11A,B)$.

Answer

132

Solution

Note that $396 = 2^2\cdot3^2\cdot11$, so exactly one of A or B is divisible by 11. Since $\gcd(B,C) = $

Copyright © ARETEEM INSTITUTE. All rights reserved.

$33 = 3 \cdot 11$, it must be the case that B is divisible by 11. Therefore, $\gcd(11A, B) = 11 \times \gcd(A, B) = 132$.

Problem 2.10 Suppose that A has 9 divisors and B has 4 divisors. Find $A + B$ if $\gcd(A, B) = 7$ and $\text{lcm}(A, B) = 2205$.

Answer

476

Solution

Since A has 9 divisors, it must be of the form p^8 or $p^2 \cdot q^2$ for primes p, q. Since $\text{lcm}(A, B) = 3^2 \cdot 5 \cdot 7^2$, we must have $A = 3^2 \cdot 7^2 = 441$. Since $5 \nmid A$ it must be the case that $5 \mid B$, and since $\gcd(A, B) = 7$, $7 \mid B$. Since B has 4 factors, this implies that $B = 5 \cdot 7 = 35$. Thus $A + B = 476$.

Copyright © Areteem Institute. All rights reserved.

3   Solutions to Chapter 3 Examples

Problem 3.1 A group of pirates went to hunt for treasure. They found a chest of gold coins. They tried to equally divide the coins, but 5 coins were left over. So they picked one pirate among themselves and threw him overboard. Then they tried to divide the coins again, but now 10 coins were left over! If the chest held 100 coins, how many pirates were there originally?

Answer

19

Solution

$100 - 5 = 95$, so the number of pirates originally was a factor of 95. $100 - 10 = 90$, so after throwing one pirate overboard, the remaining number of pirates is a factor of 90.

We know $95 = 5 \times 19$ so originally there were 1, 5, 19, or 95 pirates. 1 or 5 pirates doesn't make sense (as then there wouldn't be 5 coins left over), so after throwing one pirate overboard there must be 18 or 94 pirates. Only 18 is a factor of 90, so there must have been 19 pirates originally.

Problem 3.2 Everyday Problems with Remainders

(a) Suppose that the date is Saturday March 26th. What day of the week will March 6th be next year? (Assume next year is not a leap year.)

Answer

Sunday.

Solution

Note that every 7 days is the same day of the week. Since there are 365 days in a year and $365 = 7 \cdot 52 + 1$, March 6th next year is a Sunday.

(b) Suppose it is 9 o'clock now. What time will it be 100 hours from now, if we ignore am/pm?

Answer

1.

Copyright © ARETEEM INSTITUTE. All rights reserved.

Solution

Note that every 12 hours is the same time (with a switched am/pm we don't care about). Since $100 = 12 \cdot 8 + 4$, it will be 1 o'clock 100 hours from now.

(c) What might be a better way to think of a $1000°$ angle?

Answer

$280°$ or $-80°$.

Solution

Note that a full circle is $360°$. Since $1000 = 360 \cdot 2 + 280$, we can visualize a $1000°$ angle as the same as a $280°$ angle. Alternatively, we could also think of it as a $-80°$ angle (since $1000 = 360 \cdot 3 - 80$.

Problem 3.3 Patterns!

(a) Find the units digit of 2^{2018}.

Answer

4

Solution

The units digit of the powers of 2 follow a pattern: $2, 4, 8, 6, 2, 4, 8, 6, \ldots$. The length of the cycle is 4. The exponent 2018 has remainder 2 when divided by 4, so the units digit of 2^{2018} is the second number in the cycle: 4.

(b) Find the remainder when 2^{2018} is divided by 9.

Answer

4

Solution

The remainders when dividing by 9 of the powers of 2 follow a pattern:

$$2, 4, 8, 7, 5, 1, 2, 4, 8, 7, 5, 1, \ldots.$$

Copyright © ARETEEM INSTITUTE. All rights reserved.

The exponent 2018 has remainder 2 when divided by 4, so the units digit of 2^{2018} is the second number in the cycle: 4.

Problem 3.4 Prove the equivalence mentioned in the beginning of the packet: $m \mid (a-b)$ if and only if a and b have the same remainder when divided by m.

Solution

By the division algorithm, we can write $a = qm + r$ and $b = pm + s$ where q, p are quotients and r, s are remainders. Therefore $a - b = (q - p)m + (r - s)$. Therefore, $m \mid (a-b)$ if and only if $m \mid r - s$. If $r = s$, then $r - s = 0$ and $m \mid r - s$ as needed. Since $0 \leq r, s < m$ we have $-m < r - s < m$, so if $m \mid (r - s)$ it must be the case that $r - s = 0$, or $r = s$. This completes the proof.

Problem 3.5 Assume that $a \equiv b \pmod{m}$ and $c \equiv d \pmod{m}$. Are the following true or false? If false, come up with a counterexample. If true, you'll prove it on your homework!

(a) If $b \equiv c \pmod{m}$ then $a \equiv c \pmod{m}$.

Solution

True.

(b) $(a + c) \equiv (b + d) \pmod{m}$.

Solution

True.

(c) $(a \cdot c) \equiv (b \cdot d) \pmod{m}$.

Solution

True.

(d) If k is an integer and $k \mid a, k \mid b$, then $(a/k) \equiv (b/k) \pmod{m}$.

Answer

False

Copyright © ARETEEM INSTITUTE. All rights reserved.

Solution

Let $a = 10, b = 14, m = 4$ (so $a \equiv b \pmod 4$)). However, if $k = 2$, then $a/k = 5, b/k = 7$ and $5 \not\equiv 7 \pmod 4$.

(e) If n is a positive integer, then $a^n \equiv b^n$.

Solution

True.

Problem 3.6 Find the remainder when

(a) $35^{53} + 53^{35}$ is divided by 10.

Answer

2

Solution

The units digit of 35^{53} is 5. The units digit of 53^{35} follows the pattern $3, 9, 7, 1, 3, 9, 7, 1, \ldots$. As $35 = 8 \cdot 4 + 3$, this means the last digit of 53^{35} is 7. The sum of these last digits is $5 + 7 = 12$, whose units digit is 2.

(b) $31^{2018} + 33^{2018}$ is divided by 32.

Answer

2.

Solution

Note that $31 \equiv -1 \pmod 8, 33 \equiv 1 \pmod 8$, so $31^{2018} + 33^{2018} \equiv (-1)^{2018} + 1^{2018} \equiv 1 + 1 \equiv 2 \pmod{32}$, so the remainder is 2.

Problem 3.7 If $m > 1$ and $60 \equiv 70 \equiv 85 \pmod m$, what is m?

Answer

5

Copyright © ARETEEM INSTITUTE. All rights reserved.

Solution

Note $70 - 60 = 10, 85 - 70 = 15$, so $m \mid \gcd(10, 15) = 5$. Since $m > 1$ (and 5 is prime) m must be 5.

Problem 3.8 Suppose A, B, C, D are 4 consecutive natural numbers.

(a) Find the remainder when $A + B + C + D$ is divided by 4.

Answer

2.

Solution

Note in some order, A, B, C, D have remainders $0, 1, 2, 3$ when divided by 4. The sum therefore has a remainder of $0 + 1 + 2 + 3 \equiv 2 \pmod{4}$.

(b) Suppose you also know that $A + B + C + D$ is a three-digit number between 400 and 440 and $A + B + C + D$ is divisible by 9. Find A, B, C, D.

Answer

102, 103, 104, 105.

Solution

We know by part (a) the sum of the four natural numbers is equivalent to 2 (mod 4). Between 400 and 440, the possibilities are: 402, 406, 410, 414, 418, 422, 426, 430, 434, and 438. The only one that is a multiple of 9 is 414. The average of the four numbers is $414/4 = 103.5$, so the four numbers must be $102, 103, 104, 105$.

Problem 3.9 Let $n = \overline{a_k a_{k-1} \ldots a_1 a_0} = a_k 10^k + a_{k-1} 10^{k-1} + \cdots + a_1 10 + a_0, a_i \in \{0, 1, \ldots, 9\}$. Prove the following. Note: These are similar to things you've already proven, but practice using modular arithmetic here!

(a) Prove that $n \equiv \overline{a_{j-1} a_{j-2} \ldots a_1 a_0} \pmod{2^j}$.

Solution

Since $2^j \mid 10^j$, we have (using the properties of modular arithmetic proved above) $n =$

Copyright © ARETEEM INSTITUTE. All rights reserved.

$a_k 10^k + a_{k-1} 10^{k-1} + \cdots + a_1 10 + a_0 \equiv a_{j-1} 10^{j-1} + a_{j-2} 10^{j-2} + \cdots + a_0 \equiv \overline{a_{j-1} a_{j-2} \ldots a_1 a_0}$ (mod 2^j).

(b) Prove that $n \equiv (a_k + a_{k-1} + \cdots + a_1 + a_0)$ (mod 9). Note this means that a number is equal to the sum of its digits modulo 9.

Solution

Since $10 \equiv 1$ (mod 9), we have (again using the properties of modular arithmetic proved above) $n = a_k 10^k + a_{k-1} 10^{k-1} + \cdots + a_1 10 + a_0 \equiv a_k 1^k + a_{k-1} 1^{k-1} + \cdots + a_0 \equiv a_k + a_{k-1} + \cdots + a_0$ (mod 9).

Problem 3.10 Concatenate the positive integers $1, 2, 3, \ldots, 2017$ to form a new integer:

$$1234567891011121314 \cdots 201520162017.$$

What is the remainder when this new integer is divided by 9?

Answer

1

Solution

We want to find the number modulo 9. Modulo 9, it is equivalent to the sum of the digits, which is equivalent to the sum $1 + 2 + 3 + \cdots + 2017$, whose result is $\dfrac{2017 \times 2018}{2} = 2017 \times 1009$. We then have $2017 \equiv 2 + 0 + 1 + 7 \equiv 1$ (mod 9) and similarly $1009 \equiv 1$ (mod 9). Hence our answer is $1 \cdot 1 = 1$.

Copyright © ARETEEM INSTITUTE. All rights reserved.

4 Solutions to Chapter 4 Examples

Problem 4.1 For each of the following descriptions of a sequence: (i) write out the first few terms of the sequence, (ii) state whether it is an infinite or finite sequence, (iii) if it is a finite sequence, state its length.

(a) The number of days in the month, starting with January, February, etc.

Answer

$31, 28, 31, 30, \ldots$

Solution

Since January has 31 days, the first term of the sequence is 31. Since February has 28 days, the second term of the sequence is 28. We continue the sequence as follows:

$$31, 28, 31, 30, \ldots.$$

Since there are finitely many months in a year, the sequence is finite with length 12.

(b) The sequence with where the nth term is $4n^2 + 2$.

Answer

$2, 6, 18, 38, \ldots$

Solution

The first few terms of the sequence is

$$2, 6, 18, 38, \ldots$$

Since there is no restriction on the value of n, the sequence is infinite.

(c) The sequence where the nth term is the remainder when 2^n is divided by 9.

Answer

$2, 4, 8, 7, 5, 1, \ldots$

Copyright © Areteem Institute. All rights reserved.

Solution

The first few terms of the sequence is

$$2, 4, 8, 7, 5, 1, \ldots$$

This sequence is infinite, but in fact will start repeating the above values over and over.

Problem 4.2 Arithmetic Sequences

(a) A sequence is given with the formula $a_n = 8 - 7n$. Write out the first 5 terms of this sequence.

Answer

$8, 1, -6, -13, -20$

Solution

Directly calculating the first 5 terms are 8, $8 - 7(1) = 1$, $8 - 7(2) = -6$, $8 - 7(3) = -13$, and $8 - 7(4) = -20$.

(b) A sequence starts with the terms $-4, 8, 20, 32, 44, \ldots$. Find a general equation for a_n. Remember $a_0 = -4$.

Answer

$a_n = -4 + 12n$

Solution

The first term is -4. Each successive term we add 12. This gives an equation of $a_n = -4 + 12n$.

Problem 4.3 Geometric Sequences

(a) A sequence is given with the formula $a_n = 2 \cdot (-2)^n$. Write out the first 5 terms of this sequence.

Answer

$2, -4, 8, -16, 32$

Copyright © ARETEEM INSTITUTE. All rights reserved.

Solution

Directly calculating the first 5 terms are $2 \cdot 1 = 2, 2 \cdot (-2) = -4, 2 \cdot 4 = 8, 2 \cdot (-8) = -16$, and $2 \cdot 16 = 32$.

(b) A sequence starts with the terms $625, 125, 25, 5, 1, \ldots$. Find a general equation for a_n. Remember $a_0 = 625$.

Answer

$$a_n = 625 \cdot \left(\frac{1}{5}\right)^n$$

Solution

The first term is 625 and each successive term is $\dfrac{1}{5}$ of the previous term. Therefore the general equation is $a_n = 625 \cdot \left(\frac{1}{5}\right)^n$.

Problem 4.4 For each of the following sequences, find the first few differences. Do you notice any patterns?

(a) Arithmetic Sequence: $a_n = 3 + 5n$

Answer

The second differences are all 0

Solution

The first few terms of the sequence are $3, 8, 13, 18, 23, 31, \ldots$ which gives first differences of $5, 5, 5, 5, 5, \ldots$. Therefore the second differences are all 0: $0, 0, 0, 0, \ldots$.

(b) Geometric Sequence: $a_n = 2 \cdot 2^n$

Answer

The differences are all equal to the original sequence

Solution

The first few terms of the sequence are $2, 4, 8, 16, 32, 64, \ldots$. This gives first differences

Copyright © ARETEEM INSTITUTE. All rights reserved.

of $2, 4, 8, 16, 32, \ldots$ and second differences of $2, 4, 8, 16, \ldots$. In fact, we see that all the differences are the same as the original sequence.

(c) Cubic Sequence: $a_n = n^3 + 1$

Answer

The fourth differences are all 0

Solution

Writing out the first few terms of the sequence gives $1, 2, 9, 28, 65, 126, \ldots$ so the first differences are $1, 7, 19, 37, 61, \ldots$. Thus the second differences are $6, 12, 18, 24, \ldots$ so the third differences are $6, 6, 6, \ldots$. Hence the fourth differences, $0, 0, \ldots$, are all 0.

Problem 4.5 Consider the sequence $2, 0, 0, 2, 6, 12, 20, \ldots$. Use the method of differences, prove that this sequence has a quadratic equation and find the equation.

Answer

$a_n = n^2 - 3n + 2$

Solution

The first differences are $-2, 0, 2, 4, 6, 8, \ldots$, so the second differences are $2, 2, 2, 2, 2, \ldots$ and hence the third differences are all 0.

As the third differences are all 0, the sequence has a second degree equation: $a_n = An^2 + Bn + C$. Since $a_0 = 2$ we have $C = 2$. Using $a_1 = 0$ and $a_2 = 0$ we get that

$$A + B + 2 = 0 \text{ and } 4A + 2B + 2 = 0.$$

Doubling the first equation gives $2A + 2B + 4 = 0$ and subtracting this from the second gives $2A - 2 = 0$ so $A = 1$. Hence $1 + B + 2 = 0$ so $B = -3$. Therefore $a_n = n^2 - 3n + 2$.

Problem 4.6 Summation Notation Introduction

(a) Write out the terms of the series $\sum_{n=2}^{6} 3n^2$ and calculate the sum.

Answer

270

Copyright © ARETEEM INSTITUTE. All rights reserved.

Solution

We have

$$\sum_{n=2}^{6} 3n^2 = 3\cdot 2^2 + 3\cdot 3^2 + 3\cdot 4^2 + 3\cdot 5^2 + 3\cdot 6^2$$

$$= 12 + 27 + 48 + 75 + 108 = 270$$

(b) Write out the terms of the series $\sum_{n=1}^{5} (-1)^n \dfrac{1}{n}$ and calculate the sum.

Answer

$-\dfrac{47}{60}$

Solution

We have

$$\sum_{n=1}^{5} (-1)^n \frac{1}{n} = -1 + \frac{1}{2} - \frac{1}{3} + \frac{1}{4} - \frac{1}{5} = \frac{-47}{60}$$

(c) Find the sum $2 + 9 + 28 + 65 + 126$. What is this in summation notation?

Answer

$\sum_{n=1}^{5} n^3 + 1 = 230$

Solution

Adding we have that the sum is 230. Noting that the terms in the sequence follow the equation $n^3 + 1$ for $n = 1$ up to $n = 5$ this gives us $\sum_{n=1}^{5} n^3 + 1 = 230$ using summation notation.

(d) Find the sum $-3 + 9 - 27 + 81 - 243 + 729 - 2187$. What is this in summation notation?

Answer

$\sum_{n=1}^{7} \cdot (-3)^n = -1641$

Copyright © Areteem Institute. All rights reserved.

Solution

Calculating the sum we have -1641. Note that the terms are given by the equation $(-3)^n$ for $n = 1$ up to $n = 7$ we have that $-1641 = \displaystyle\sum_{n=1}^{7} (-3)^n$ in summation notation.

Problem 4.7 Arithmetic Series

(a) Calculate $\displaystyle\sum_{k=1}^{n} k$.

Answer

$\dfrac{n(n+1)}{2}$

Solution

We want to calculate the sum $1 + 2 + 3 + \cdots + n$. Notice writing the sum forwards and backwards and adding we have

$$
\begin{array}{ccccccccc}
 & 1 & + & 2 & + & 3 & + \cdots + & n \\
+ & n & + & n-1 & + & n-2 & + \cdots + & 1 \\
\hline
= & (n+1) & + & (n+1) & + & (n+1) & + \cdots + & (n+1)
\end{array}
$$

which equals $n \cdot (n+1)$. As this is double the sum we want we have $1 + 2 + \cdots + n = \dfrac{n(n+1)}{2}$.

(b) Calculate the sum of the first 20 terms of the sequence $7, 5, 3, 1, -1, \ldots$.

Answer

-240

Solution

The sequence has equation $a_n = 7 - 2n$ and hence $a_{19} = 7 - 2(19) = -31$ so we want to calculate the sum $7 + 5 + 3 + \cdots + -31$. Notice writing the sum forwards and backwards and adding we have

$$
\begin{array}{ccccccccc}
 & 7 & + & 5 & + & 3 & + \cdots + & -31 \\
+ & -31 & + & -29 & + & -27 & + \cdots + & 7 \\
\hline
= & -24 & + & -24 & + & -24 & + \cdots + & -24
\end{array}
$$

Copyright © ARETEEM INSTITUTE. All rights reserved.

which equals $-24 \cdot 20 = -480$. As this is double the sum we want we have $7 + 5 + 3 + \cdots + -31 = -480 \div 2 = -240$.

(c) Suppose $a_n = 3n - 2$. Calculate $\displaystyle\sum_{k=2}^{18} a_k$.

Answer

476

Solution

$a_2 = 3(2) - 2 = 4$ and $a_{18} = 3(18) - 2 = 52$ so we want to calculate the sum $4 + 7 + 10 + \cdots + 52$. Notice writing the sum forwards and backwards and adding we have

$$
\begin{array}{ccccccccc}
 & 4 & + & 7 & + & 10 & + & \cdots & + & 52 \\
+ & 52 & + & 49 & + & 46 & + & \cdots & + & 4 \\
\hline
= & 56 & + & 56 & + & 56 & + & \cdots & + & 56
\end{array}
$$

which equals $56 \cdot (18 - 2 + 1) = 952$. As this is double the sum we want we have $4 + 7 + 10 + \cdots + 52 = 952 \div 2 = 476$.

Problem 4.8 The repeating decimal $0.\overline{9}$

(a) Write the repeating fraction $0.\overline{9}$ as an (infinite) geometric series.

Answer

$$0.\overline{9} = \sum_{k=1}^{\infty} 9 \cdot \left(\frac{1}{10}\right)^k$$

Solution

We have the decimal

$$0.\overline{9} = 0.9 + 0.09 + 0.009 + \cdots = 9 \cdot \frac{1}{10} + 9 \cdot \frac{1}{100} + 9 \cdot \frac{1}{1000} + \cdots$$

so using summation notation we have $0.\overline{9} = \displaystyle\sum_{k=1}^{\infty} 9 \cdot \left(\frac{1}{10}\right)^k$.

(b) Let $0.\overline{9} = x$. Consider the expression $10x - x$ and use this expression to solve for x.

Copyright © AReteem Institute. All rights reserved.

Answer

$x = 1$

Solution

If $0.\overline{9} = x$ we have $10x = 9.\overline{9}$. Hence $10x - x = 9.\overline{9} - 0.\overline{9}$ and $9x = 9$. Thus $x = 1$.

Problem 4.9 Geometric Series

(a) Calculate the sums $\displaystyle\sum_{k=0}^{n} 2^k$ for $n = 1, 2, 3, 4, 5$. Do you notice a pattern?

Answer

$$\sum_{k=0}^{n} 2^k = 2^{n+1} - 1$$

Solution

The powers of 2 are $1, 2, 4, 8, 16, 32$. Thus our sums are

$1, 1+2 = 3, 1+2+4 = 7, 1+2+4+8 = 15, 1+2+4+6+16 = 31$, and $1+2+4+8+16+32 = 63$.

Note that each of these is one less than the next power of 2, so we guess $\displaystyle\sum_{k=0}^{n} 2^k = 2^{n+1} - 1$.

(b) Using a method similar to the previous problem, explain a general formula for $\displaystyle\sum_{k=0}^{n} 2^k$.

Answer

$$\sum_{k=0}^{n} 2^k = 2^{n+1} - 1$$

Solution

Set

$$
\begin{aligned}
S &= 1 + 2 + 4 + \cdots + 2^n \\
\text{so} \quad 2S &= 2 + 4 + 8 + \cdots + 2^{n+1}
\end{aligned}
$$

Note when subtracting almost all the terms cancel so we have $S = 2S - S = 2^{n+1} - 1$ which gives a general formula for the sum.

Copyright © ARETEEM INSTITUTE. All rights reserved.

(c) For what values of $r \neq \pm 1$ in a geometric sequence $a_n = a_0 \cdot r^n$ does it make sense to have the infinite series $\sum\limits_{k=0}^{\infty} a_k$?

Answer

$|r| < 1$

Solution

Clearly if $|r| > 1$ when the sequence is getting larger and larger, so the infinite series does not make sense. In face when $|r| < 1$ we can use a method similar to the previous problem to find a general formula.

The case when $|r| = 1$ is more complicated, and we won't cover the details in this class. In fact the infinite series makes sense when $r = -1$ but not when $r = 1$.

(d) Calculate $\sum\limits_{k=0}^{\infty} 9 \cdot \left(\dfrac{1}{3}\right)^k$.

Answer

$\dfrac{27}{2}$

Solution

Set

$$\begin{aligned} S &= 9 + 3 + 1 + \tfrac{1}{3} + \cdots \\ \text{so} \quad \tfrac{1}{3}S &= 3 + 1 + \tfrac{1}{3} + \tfrac{1}{9} + \cdots \end{aligned}$$

Note when subtracting almost all the terms cancel so we have

$$\frac{2}{3}S = S - \frac{1}{3}S = 9 \Rightarrow S = \frac{3}{2} \cdot 9 = \frac{27}{2}$$

as the sum of the infinite series.

Problem 4.10 What is the units digit of $1^2 + 2^2 + 3^2 + \cdots + 99^2$?

Answer

0

Copyright © ARETEEM INSTITUTE. All rights reserved.

Solution

Note that $\overline{a0}^2 + \overline{a1}^2 + \overline{a2}^2 + \cdots + \overline{a9}^2 \equiv 1^2 + 2^2 + \cdots + 9^2 \pmod{10}$ (for example, $20^2 + 21^2 + \cdots 29^2 = 0^2 + 1^2 + \cdots + 9^2$). Hence, $1^2 + 2^2 + \cdots 99^2 \equiv 10 \times (1^2 + 2^2 + \cdots + 9^2) \equiv 0 \pmod{10}$ so the units digit is 0.

Alternatively, we have $1^2 + 2^2 + \cdots + 99^2 = \dfrac{99 \times 100 \times 199}{6} = 328350$, so the units digit is 0.

Copyright © ARETEEM INSTITUTE. All rights reserved.

5 Solutions to Chapter 5 Examples

Problem 5.1 List the first few terms of each sequence and verify that both formulas lead to the same sequence:

$a_n = 5 + 3n$ and $b_0 = 5$, $b_{n+1} = b_n + 3$

Solution

Using either equation we see the first the first few terms are $5, 8, 11, 14, 17, 20, \ldots$.

Problem 5.2 List the first few terms of each sequence and verify that both formulas lead to the same sequence:

$a_n = 2 \cdot 3^n$ and $b_0 = 2$, $b_{n+1} = b_n \cdot 3$

Answer

Solution

Using either equation we see the first the first few terms are $2, 6, 18, 54, 162, 486, \ldots$

Problem 5.3 Arithmetic sequences are given below, in one of three ways: (i) the first few terms of the sequence, (ii) the formula, or (iii) the recursive formula. Give the other 2 ways of describing the sequence. (That is, if the recursive formula is given, write out the first few terms and give the general formula for the sequence.)

(a) $a_0 = 5$, $a_{n+1} = 8 + a_n$.

Answer

$5, 13, 21, 29, \ldots$; $a_n = 5 + 8n$

Solution

We are given the recursive definition. The first few terms are $5, 13, 21, 29, \ldots$, starting with 5 and adding 8 each time. Hence the formula is $a_n = 5 + 8n$.

(b) $3, 5, 7, \ldots$.

Copyright © ARETEEM INSTITUTE. All rights reserved.

Answer

$a_0 = 3, a_{n+1} = a_n + 2$ or $a_n = 3 + 2n$

Solution

We are given the first few terms, starting with 3 and adding 2 each time. Hence the recursive formula is $a_0 = 3, a_{n+1} = a_n + 2$ and the formula is $a_n = 5 + 8n$.

(c) $a_n = 6 - 5n$.

Answer

$6, 1, -4, -9, \ldots; a_0 = 6, a_{n+1} = a_n - 5$

Solution

We are given the general equation. The first few terms are $6, 1, -4, -9, \ldots$, starting with 6 and subtracting 5 each time. Hence the recursive formula is $a_0 = 6, a_{n+1} = 6 - 5n$.

Problem 5.4 Geometric sequences are given below, in one of three ways: (i) the first few terms of the sequence, (ii) the formula, or (iii) the recursive formula. Give the other 2 ways of describing the sequence. (That is, if the recursive formula is given, write out the first few terms and give the general formula for the sequence.)

(a) $2, 4, 8, \ldots$

Answer

$a_0 = 2, a_{n+1} = 2 \cdot a_n; a_n = 2 \cdot 2^n$

Solution

We are given the first few terms, starting with 2 and multiplying by 2 each time. Hence the recursive formula is $a_0 = 2, a_{n+1} = 2 \cdot a_n$ and the formula is $a_n = 2 \cdot 2^n$.

(b) $a_0 = -3, a_{n+1} = -2 \cdot a_n$.

Answer

$-3, 6, -12, \ldots; a_n = -3 \cdot (-2)^n$

Copyright © Areteem Institute. All rights reserved.

Solution

We are given the recursive definition. The first few terms are $-3, 6, -12, \ldots$, starting with -3 and multiplying by -2 each time. Hence the formula is $a_n = -3 \cdot (-2)^n$.

(c) $a_n = 4^n$.

Answer

$1, 4, 16, \ldots$; $a_0 = 1, a_{n+1} = 4 \cdot a_n$

Solution

We are given the general equation. The first few terms are $1, 4, 16, \ldots$, starting with 1 and multiplying by 4 each time. Hence the recursive formula is $a_0 = 1, a_{n+1} = 4 \cdot a_n$.

Problem 5.5 Verify algebraically that the formulas and recursive formulas lead to the same sequence

(a) $a_n = 5 + 4n$; $a_0 = 5$, $a_{n+1} = a_n + 4$.

Solution

First we have $a_0 = 5 + 4(0) = 5$ as needed. If $a_n = 5 + 4n$ then

$$a_n + 4 = (5 + 4n) + 4 = 5 + 4(n+1) = a_{n+1}$$

as given in the recursive formula. (Here we are plugging in $(n+1)$ for n in the general equation.)

(b) $a_n = 3 \cdot 2^n$; $a_0 = 3$, $a_{n+1} = 2 \cdot a_n$.

Solution

First we have $a_0 = 3 \cdot 2^0 = 3$ as needed. If $a_n = 3 \cdot 2^n$ then

$$2 \cdot a_n = 2 \cdot (3 \cdot 2^n) = 3 \cdot 2^{n+1} = a_{n+1}$$

as given in the recursive formula. (Here we are plugging in $(n+1)$ for n in the general equation.)

Problem 5.6 Suppose a sequence starts $G_0 = 2$, $G_1 = 1$, $G_{n+1} = 2 \times G_n - G_{n-1}$. That

Copyright © ARETEEM INSTITUTE. All rights reserved.

is, multiply the previous term by 2 and subtract the term before that. Is there a simpler formula for this sequence?

Answer

Yes, $G_n = 2 - n$

Solution

Given that the first two terms of the sequence is 2 and 1, we see that the third term of the sequence is
$$2 \times 1 - 2 = 0.$$
Therefore, the fourth term of the sequence is
$$2 \times 0 - 1 = -1.$$
Repeating the above procedure yields that the next 5 terms of the sequence is,
$$0, -1, -2, -3, -4.$$
Thus G_n starts with 2 and decreases by 1 each term, so a simpler formula for G_n is $G_n = 2 - n$.

Problem 5.7 Review of common differences

(a) Use common differences to determine what kind of formula the recursive sequence $a_3 = 0$, $a_{n+1} = a_n + n - 1$ has.

Answer

Quadratic: $a_n = An^2 + Bn + C$

Solution

The first few terms of the sequence (starting with 3) are $0, 2, 5, 9, 14, \ldots$ so the first differences are $2, 3, 4, 5, \ldots$. Thus the second differences are $1, 1, 1, \ldots$ and the third differences are all 0. This means the sequence is given by a $3 - 1 = 2$ degree polynomial, a quadratic with the form $An^2 + Bn + C$.

(b) Find a formula for the sequence.

Answer

$$a_n = \frac{n(n-3)}{2}$$

Copyright © ARETEEM INSTITUTE. All rights reserved.

Solution

We know that $a_3 = 0, a_4 = 2, a_5 = 5$ and by part (a) $a_n = An^2 + Bn + C$. Plugging in $n = 3, 4, 5$ we have

$$a_3 = 9A + 3B + C = 0, a_4 = 16A + 4B + C = 2, a_5 = 25A + 5B + C = 5$$

Looking at $a_4 - a_3$ and $a_5 - a_4$ we get that

$$7A + B = 2 \text{ and } 9A + B = 3.$$

Subtracting these again we get that $2A = 1$ so $A = \dfrac{1}{2}$. Hence $B = 2 - 7A = \dfrac{-3}{2}$. Lastly $C = -9A - 3B = 0$. Therefore $a_n = \dfrac{1}{2}n^2 - \dfrac{3}{2}n = \dfrac{n(n-3)}{2}$.

Problem 5.8 Find a formula for the sequence defined by the recursive formula $a_1 = 2$, $a_{n+1} = 3a_n + 2$

Answer

$a_n = 3^n - 1$

Solution

Notice that the recursive formula includes $3 \cdot a_n$ as part of the equation. Because of this we guess that a_n has a general formula of the form $a_n = A \cdot 3^n + B$ for numbers A, B. We know $a_1 = 2$ so $A \cdot 3^1 + B = 2$ or $3A + B = 2$. $a_2 = 3 \cdot 2 + 2 = 8$ so $A \cdot 3^2 + B = 8$ or $9A + B = 8$. Subtracting we get $6A = 6$ or $A = 1$. Then $B = 2 - 3A = -1$. Therefore $a_n = 3^n - 1$.

Problem 5.9 Recall the Fibonacci sequence defined by $F_1 = F_2 = 1$, and $F_{n+2} = F_{n+1} + F_n$. For $n \geq 1$, let S_n be the sum of the first n terms of the Fibonacci sequence. ($S_n = F_1 + \cdots + F_n$ or $S_1 = F_1$, $S_2 = F_1 + F_2$, etc.)

(a) Write out S_n for $n = 1, 2, 3, 4, 5$.

Answer

$1, 2, 4, 7, 12$

Solution

We know the first 5 terms of the Fibonacci sequence is $1, 1, 2, 3, 5$. Therefore the first

Copyright © ARETEEM INSTITUTE. All rights reserved.

few terms of S_n are

$$1, 1+1 = 2, 1+1+2 = 4, 1+1+2+3 = 7, \text{ and } 1+1+2+3+5 = 12$$

(b) Compare your answer in (a) with the original Fibonacci sequence.

Answer

$S_n = F_{n+2} - 1$

Solution

In (a) we calculated the first 5 terms of S_n as $1, 2, 4, 7, 12, \ldots$. For reference we again look at the Fibonacci sequence (with a few more terms): $1, 1, 2, 3, 5, 8, 13, 21, \ldots$. We notice that $12 = 13 - 1$ and $7 = 8 - 1$. In fact this pattern continues, with $4 = 5 - 1$, $2 = 3 - 1$, and $1 = 2 - 1$ so in fact $S_n = F_{n+2} - 1$.

Problem 5.10 More on recursive sequences

(a) Define a sequence generated by the following: start with 12 and divide by 2 if the number is even or take 3 times the number plus 1 if the number is odd. What is the largest value a term of this sequence can have?

Answer

16

Solution

Since the first term is even, the second term is $12 \div 2 = 6$. Since the second term is even, the third term is $6 \div 2 = 3$. Since the third term is odd, the fourth term is $3 \times 3 + 1 = 10$.

Implementing the pattern, we get:

$$12, 6, 3, 10, 5, 16, 8, 4, 2, 1, 4, 2, 1, \ldots$$

where the pattern keeps repeating $4, 2, 1$ from then on. Hence the largest value in the sequence is 16.

(b) What is the 8th term in the sequence 1, 11, 21, 1211, 111221, 312211?

Copyright © ARETEEM INSTITUTE. All rights reserved.

Answer

1113213211

Solution

Let us think of reading the terms out loud, one digit at a time. The first is one, the second is one one, the third is two one, and the fourth is one two one one.

What did we say for the first term? Well we said one once, or one one. The second time we said one twice, or two one. This pattern continues, with the third time being one two then one one or one two one one. Continuing this pattern the first 8 terms of the sequence are

$$1, 11, 21, 1211, 111221, 312211, 13112221, 1113213211$$

so our answer is 1113213211.

Copyright © ARETEEM INSTITUTE. All rights reserved.

6 Solutions to Chapter 6 Examples

Problem 6.1 Suppose John has 2 hats, 5 shirts, 1 jacket, 4 pairs of pants, 3 pairs of shorts, and 4 pairs of shoes.

(a) Suppose John makes an outfit consisting of a shirt, a pair of pants, and a pair of shoes. How many different outfits does he have?

Answer

80

Solution

We use the product rule: $5 \cdot 4 \cdot 4 = 80$.

(b) Repeat (a) if John *can* wear shorts instead of pants.

Answer

140

Solution

Using the Sum Rule, we have $4 + 3 = 7$ choices for leg wear. We then proceed as in (a): $5 \cdot 7 \cdot 4 = 140$.

(c) Now suppose John can wear shorts or pants as in (b), *but* if he wears shorts, he will also wear a hat and possibly a jacket.

Answer

320

Solution

Consider the two cases (so Sum Rule) based on whether John wears shorts or pants. If he wears pants, it is the same as (a); if he wears shorts, we also have to choose which hat he wears and whether or not he wears a jacket (2 choices). The total is: $5 \cdot 4 \cdot 4 + 5 \cdot 3 \cdot 4 \cdot 2 \cdot (1 + 1) = 320$.

Copyright © ARETEEM INSTITUTE. All rights reserved.

Problem 6.2 Suppose you have a group of 6 people. How many different photographs are there of everyone lined up if:

(a) all the people look different?

Solution

$6! = 720$.

(b) 2 of the people are identical twins who have dressed identically?

Answer

$$\frac{6!}{2!} = \binom{6}{2} \cdot 4! = 360.$$

Solution

Two methods are presented above. For the first, we divide the answer from (a) by 2 because we do not care about the order of the twins (since we cannot tell the twins apart). For the second, we first choose 2 spots for the twins to stand, and then fill in the other 4 people around them.

(c) 2 of the people are a couple and must stand next to each other?

Answer

$2! \cdot 5! = 240$.

Solution

There are $2 = 2!$ ways to decide the arrangement of the couple. If we then treat them as a single 'object' we must arrange 5 'objects'.

(d) 2 of the people are sworn enemies and cannot stand next to each other?

Answer

$6! - 2! \cdot 5! = 480$.

Copyright © ARETEEM INSTITUTE. All rights reserved.

Solution

We count the total number of arrangement and subtract off the number of arrangements where they are next to each other (which is equivalent to them being a couple).

Problem 6.3 There are 30 lottery balls labeled from 1 to 30.

(a) How many ways are there to draw 5 lottery balls, in order one after another, if we do not replace the ball after each pick? (That is, it is not possible to pick the same ball more than once.)

Answer

17100720

Solution

There are 30 possible lottery balls to choose for the first drawing. After the first lottery ball is chosen, there are 29 remaining lottery balls to choose for in the second drawing. Continuing the procedure, there is a total of

$$30 \times 29 \times 28 \times 27 \times 26 = 17100720$$

ways to choose 5 lottery balls without replacement.

(b) How many ways are there to draw 5 lottery balls all at once? (That is, it is not possible to draw the same ball twice, and the 5 balls are in no particular order.)

Answer

$$\binom{30}{5} = \frac{30!}{5!25!} = 142506$$

Solution

This is equivalent to drawing 5 indistinguishable lottery balls from a pool of 40 balls. From the previous problem, there is a total of

$$30 \times 29 \times 28 \times 27 \times 26 = 17100720$$

ways to choose 5 lottery balls without replacement.

Since the lottery balls are indistinguishable, we divide the number of arrangements of the 5 balls to remove the duplicates from the above count. Therefore, there are

$$17100720 \div 5! = 142506$$

Copyright © Areteem Institute. All rights reserved.

ways to draw 5 lottery balls all at once.

Problem 6.4 How many rearrangements can be made of the letters in the word BA-NANAS?

Answer

420

Solution

Pretend that all of the letters are distinct. There are

$$7! = 7 \times 6 \times 5 \times 4 \times 3 \times 2 \times 1 = 840$$

ways to rearrange 7 letters to form distinct words.

Since there are 3 A's and 2 N's in *BANANAS*, we need to rid the duplicates by dividing the total number of ways to rearrange 7 different letters by $3! \times 2! = 6 \times 2 = 12$.

Therefore, the answer is

$$\frac{7!}{3! \cdot 2!} = \frac{7 \times 6 \times 5 \times 4 \times 3 \times 2 \times 1}{3 \times 2 \times 1 \times 2 \times 1} = 420.$$

Problem 6.5 10 friends decide to play five versus five basketball, so they need to divide themselves into two teams. If there is no order associated with how the people are picked and no order associated with the teams, how many different ways can they divide themselves into the two teams?

Answer

126

Solution

There are $\binom{10}{5} \cdot \binom{5}{5}$ ways to divide themselves into a first team and a second team. We do not care about the order of the teams, so we also divide by 2. This gives a final answer of $\binom{10}{5} \div 2 = 126$ total ways to divide themselves into teams.

Copyright © ARETEEM INSTITUTE. All rights reserved.

Problem 6.6 10 points are marked on the plane. How many different triangles can be formed using these points as vertices if no three of the points are in a straight line?

Answer

120

Solution

Note that if no three points are on a straight line we can choose any group of 3 points to form a triangle. This can be done in

$$\binom{10}{3} = 120$$

ways.

Problem 6.7 Suppose you write out the numbers $1 - 1000$: $1, 2, 3, 4, \ldots, 1000$.

(a) How many digits have you written in total?

Answer

2893

Solution

$1 \cdot 9 + 2 \cdot 90 + 3 \cdot 900 + 4 \cdot 1 = 2893$.

(b) What is the sum of all the numbers written?

Solution

$$1 + 2 + 3 + \cdot 1000 = \frac{1000 \cdot 1001}{2} = 500500.$$

(c) What is the sum of all the digits written?

Answer

$(0 + 1 + 2 + 3 + 4 + 5 + 6 + 7 + 8 + 9) \cdot 300 + 1 = 13501.$

Solution

1000 contributes a digit sum of 1. We may think of all the other numbers as containing

Copyright © ARETEEM INSTITUTE. All rights reserved.

3 digits (including leading 0's if needed) and we can also include 000 without changing the sum. Each digit is thus written a total of $3 \cdot 10^2 = 300$ times.

Problem 6.8 Suppose a pizza place has 5 toppings available. You want to order 2 different 3-topping pizzas. Suppose repeated toppings are not allowed on a single pizza, and the order of the toppings on a pizza does not matter. If you only care which two pizzas you get, how many ways are there to make the order?

Answer

45

Solution

Since the pizza has 5 toppings made available and each pizza requires 3 nonrepeated toppings, there are

$$\binom{5}{3} = 10$$

combinations of toppings that are made available for pizzas. Of the 10 possible pizzas, we wish to choose 2 pizzas for our order (without order), which can be done in

$$\binom{10}{2} = 45$$

ways.

Problem 6.9 It is time for Dennis to make a new password. He's not too creative so he decides to create a password with his name and birthday, which is May 25. He wants to use the letters of his name (in order) and the letters/symbols of his birthday (in order). That is, the password could start with either D or M; if it starts with D the next letter could be M or e, and if it starts with M the next letter could be D or a. Examples of possible passwords are MayDennis25 or DenMaynis25 or DeManyni2s5. A password of 25DennisMay or nisMay25Den is NOT allowed. How many possible passwords could Dennis make?

Answer

462

Solution

Since both the strings Dennis and May25 will appear in order, the password has length

Copyright © ARETEEM INSTITUTE. All rights reserved.

$5 + 6 = 11$ and is determined when we know which 6 places the letters in Dennis occupy. As there are 11 positions in total, there are $\binom{11}{6} = \dfrac{11!}{6!5!} = 462$ total passwords.

Problem 6.10 Carrie invites 9 of her friends for dinner. Carrie will sit with four of the friends at the first circular table, while the other 5 friends sit around the second table. If all the seats at both tables are indistinguishable, how many seating arrangements are there?

Answer

72576

Solution

First there are $\binom{9}{4}$ ways to choose which 4 friends sit with Carrie. The 4 friends then need to be arranged around the table with Carrie, which can be done in 4! ways. Similarly, the other 5 friends can be seating in $5! \div 4 = 4!$ ways around the second circular table. This gives a final answer of

$$\binom{9}{4} \cdot 4! \cdot 4! = \frac{9!}{5} = 72576.$$

Copyright © Areteem Institute. All rights reserved.

7 Solutions to Chapter 7 Examples

Problem 7.1 You roll 2 four-sided dice. Let A be the event that the first die is a 4, and B the event that the sum of the two rolls is 6.

(a) Write out the sample space Ω representing a list of all possible outcomes of rolling 2 four-sided dice. What is $n(\Omega)$?

Answer

16

Solution

The list of all possible outcomes from rolling 2 four-sided dice is given below:

$$\Omega = \{(1,1),(1,2),(1,3),(1,4),(2,1),(2,2),\ldots(4,4)\}.$$

There are 16 total outcomes, so $n(\Omega) = 16$.

(b) List the outcomes in event A. What is $n(A)$?

Answer

4

Solution

We have $A = \{(4,1),(4,2),(4,3),(4,4)\}$ so $n(A) = 4$.

(c) List the outcomes in event B. What is $n(B)$?

Answer

3

Solution

We have $B = \{(2,4),(3,3),(4,2)\}$ so $n(B) = 3$.

(d) List the outcomes in event $A \cap B$. What is $n(A \cap B)$?

Copyright © ARETEEM INSTITUTE. All rights reserved.

Answer

1

Solution

The only outcome in $A \cap B$ is $(4,2)$ so $n(A \cap B) = 1$.

Problem 7.2 How many numbers less than 1000 are divisible by 11 but not by 9?

Answer

80

Solution

$999 \div 11 = 90 \, R \, 9$ so there are 90 multiples of 11 less than 1000. We want to remove the numbers that are also multiples of 9, which is all multiples of $\mathrm{lcm}(9,11) = 99$. We have $999 \div 99 = 10 \, R \, 9$ there are 10 such numbers less than 1000. Hence our answer is $90 - 10 = 80$.

Problem 7.3 Students in Areteem Institute were asked which pets (dogs or cats) do they have. In a survey of 50 students, 10 of them answered "No pets", 30 answered "a dog" and 20 answered "a cat". How many students have both a cat and dog?

Answer

10

Solution

Given 50 students, since 10 of the students do not own any pets,

$$50 - 10 = 40$$

students owns at least one pet. Of the 40 remaining students, we wish to distribute the students to one of three regions in the following Venn Diagram:

Copyright © ARETEEM INSTITUTE. All rights reserved.

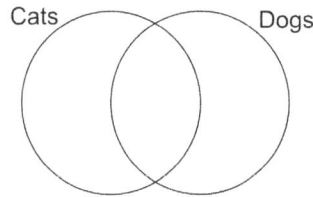

It is given that there are 20 students that answered "a cat". This indicates that the shaded regions shown below must add to 20.

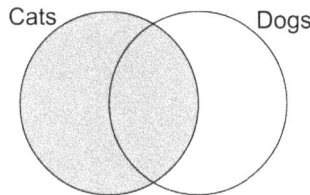

It is also given that there are 30 students that answered "a dog". This indicates that the shaded regions shown below must add to 30.

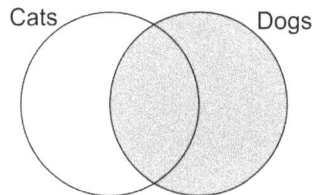

Note that if we add shaded regions above, the overlap of the regions is counted twice. This suggests that the number of students who have at least one cat and one dog is

$$20 + 30 - 40 = 10.$$

This is because the overlap of the number of students represents the number of students that own both cats and dogs. Therefore, there are 10 students that have both a cat and dog.

Copyright © ARETEEM INSTITUTE. All rights reserved.

Problem 7.4 (a) Work out and write out the PIE formula for 3 sets A, B, C.

Solution

$$n(A \cup B \cup C) = n(A) + n(B) + n(C) - n(A \cap B) - n(A \cap C) - n(B \cap C) + n(A \cap B \cap C).$$

(b) How many terms will the PIE formula for 4 sets A, B, C, D have?

Answer

15

Solution

There are $\binom{4}{1} = 4$ terms involving 1 set, $\binom{4}{2} = 6$ terms involving 2 sets, $\binom{4}{3} = 4$ involving 3 sets, and $\binom{4}{4} = 1$ with all 4 sets. Hence there are a total of $4 + 6 + 4 + 1 = 15$ terms.

Problem 7.5 How many positive integers ≤ 1000 are a perfect square, cube, fourth, or fifth power?

Answer

40

Solution

First note that any fourth power is also a square, so we just need to count how many integers are a perfect square, cube, or fifth power. Let A be the set of squares, B the set of cubes, and C the set of fifths. We want $n(A \cup B \cup C)$ and using PIE we have

$$n(A \cup B \cup C) = n(A) + n(B) + n(C) - n(A \cap B) - n(A \cap C) - n(B \cap C) + n(A \cap B \cap C).$$

Note here $A \cap B$ is the set of sixth powers, $A \cap C$ is the set of tenth powers, $B \cap C$ is the set of fifteenth powers, and $A \cap B \cap C$ is the set of thirtieth powers. From this we can see that

$$n(A) = 31, n(B) = 10, n(C) = 3, n(A \cap B) = 3, n(A \cap C) = n(B \cap C) = n(A \cap B \cap C) = 1$$

so we have

$$n(A \cup B \cup C) = 31 + 10 + 3 - 3 - 1 - 1 + 1 = 40$$

Copyright © ARETEEM INSTITUTE. All rights reserved.

as our answer.

Problem 7.6 The following functions are all between the real numbers and itself ($\mathbb{R} \to \mathbb{R}$).

For each of the functions: (i) is it injective/one-to-one?, (ii) what is its range?

(a) $y = x^3 + 3$

Answer

Injective, Range is all real numbers

Solution

Consider the graph of the function as shown below:

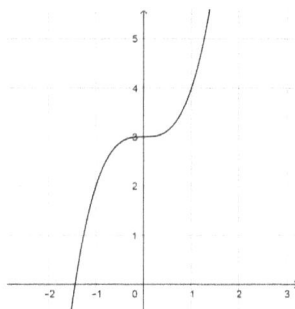

We see the function passes the horizontal line test so it is injective. We also see that the range is all real numbers.

(b) $y = x^3 - x$

Answer

Not injective, Range is all real numbers

Solution

Consider the graph of the function as shown below:

Copyright © ARETEEM INSTITUTE. All rights reserved.

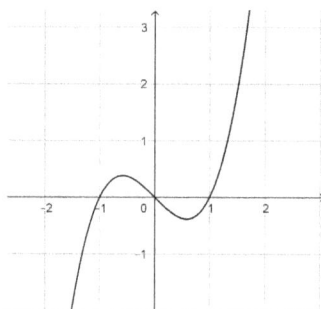

We see that when $x = 0$ or $x = \pm 1$ that $y = 0$, so the function is not injective. We can see, however, that the range is still all real numbers.

(c) $y = 2^x$

Answer

Injective, Range is all positive real numbers

Solution

Consider the graph of the function as shown below:

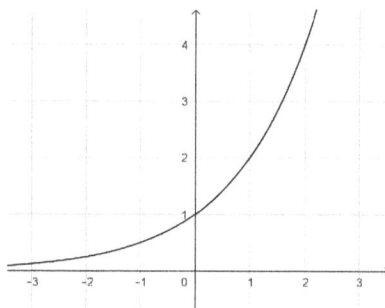

We see the function passes the horizontal line test so it is injective. However, the range is not all real numbers, it is only the positive real numbers.

Problem 7.7 Let $A = \{1,2,3,4\}$ and $B = \{1,2,3\}$. Answer each of the following. Explain how you have already answered this type of question in the past.

(a) How many total functions are there from A to B? from B to A?

Copyright © ARETEEM INSTITUTE. All rights reserved.

Answer

81, 64

Solution

For each input, we must choose one of the outputs. Hence from A to B there are $3^4 = 81$ functions while from B to A there are $4^3 = 64$ total functions.

(b) How many bijections are there from A to B?

Answer

0

Solution

A and B have different sizes, so there are no bijections between them.

(c) How many injections are there from B to A?

Answer

24

Solution

For each member in B we need a different output in A, so this is a permutation. Hence there are
$$4 \cdot 3 \cdot 2 = 24$$
injective functions.

Problem 7.8 Let $A = \{1,2,3,4\}$ and $B = \{1,2,3\}$. How many surjections are there from A to B?

Answer

36

Solution

Since A has 4 elements and B has 3, in a surjective function one of the elements of B

Copyright © ARETEEM INSTITUTE. All rights reserved.

will be the output for 2 inputs from A. There are 3 choices for which element in B is repeated twice and then $\binom{4}{2}$ ways to choose which two elements in A are mapped this this element. We are left with 2 elements in A and 2 elements in B, so there are 2! ways to complete the function. This gives $3 \cdot \binom{4}{2} \cdot 2! = 36$ surjections in all.

Problem 7.9 Suppose you have a set $S = \{1,2,3,\ldots,20\}$. You want to choose A,B,C such that $A \cup B \cup C = S$ and $A \cap B \cap C = \emptyset$. (Remember $\emptyset = \{\}$ is the empty set, which contains no elements.) How many ways an this be done

(a) if we also assume $A \cap B = A \cap C = B \cap C = \emptyset$? (Under these conditions, A,B,C *partitions S*.)

Answer

3^{20}

Solution

Every number $1,2,3,\ldots,20$ is either in A or B or C. Therefore we have 3^{20} outcomes.

(b) in total?

Answer

6^{20}

Solution

Think of the Venn diagram. Every number $1,2,3,\ldots,20$ can either be in just A, just B, just C, or in $A \cap B$, in $A \cap C$, or in $B \cap C$ (6 choices for each). Hence there are 6^{20} outcomes.

Problem 7.10 Let $S = \{1,2,3,4,5\}$. How many 5-digit numbers can be formed from members of S with no repeated digits and the 2 next to 1 or 3?

Answer

84

Copyright © ARETEEM INSTITUTE. All rights reserved.

Solution

Let A be the set of 5-digit numbers where 1 and 2 are adjacent and B the set where 2 and 3 are adjacent. Then we have

$$n(A) = n(B) = 2! \cdot 4!$$

as we group the two digits together and arrange them with the other 3, giving 4! orders times 2! for flipping the two adjacent numbers.

However, we are double counting the 5-digit numbers where 2 is adjacent to 1 and 3 (this is the set $A \cap B$). In this case we group all 3 (so there are 3 groups to arrange) and then note that since the 2 must be in the middle, there are still 2 ways to arrange the group (123 or 321). Hence $n(A \cap B) = 2! \cdot 3!$. This gives a final answer of

$$2! \cdot 4! + 2! \cdot 4! - 2! \cdot 3! = 84$$

numbers in total.

Copyright © ARETEEM INSTITUTE. All rights reserved.

8   Solutions to Chapter 8 Examples

Problem 8.1 Suppose below is a map of a city you want to travel from A to B.

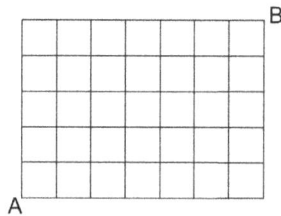

(a) If each square in the diagram is a square block what is the minimum number of blocks it takes to get from A to B?

Answer

12

Solution

In the diagram above, we are required to travel 5 blocks up and 7 blocks right to get from A to B. Therefore, the minimum number of blocks it takes to get from A to B is $5 + 7 = 12$.

(b) How many paths of shortest length are there from A to B?

Answer

792

Solution

Let R stand for right and U stand for up. Note that any one of the shortest paths consists of 5 blocks up (5 U's) and 7 blocks right (7 R's). A unique path is determined by the arrangement of 3 U's and 4 R's. The number of possible ways to arrange 5 U's and 7 R's is

$$\binom{12}{5} = \frac{12!}{5! \times 7!} = 792,$$

the number of shortest paths.

Copyright © ARETEEM INSTITUTE. All rights reserved.

Problem 8.2 More Paths

(a) Consider the grid below.

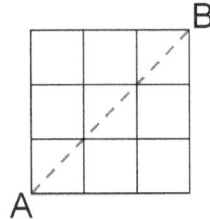

Consider paths from A to B where each step is either right (R) or up (U). How many such paths do not go below the dotted line? List out these paths.

Answer

5: $UUURRR, UURURR, UURRUR, URUURR, URURUR$

Solution

We know the first step will be U. If we start UUU we must finish as RRR giving $UUURRR$. If we start UUR we can finish URR or RUR giving $UURURR$ and $UURRUR$. Finally if we start UR we can finish $UURR$ or $URUR$ giving $URUURR$ and $URURUR$. In total we have $1 + 2 + 2 = 5$ paths that do not go below the dotted line.

(b) Consider the set of all words made up of 3 R's and 3 U's. Describe the properties one of these words should have to correspond to one of the paths as in part (a).

Solution

Looking at part (a), the paths were $UUURRR, UURURR, UURRUR, URUURR, URURUR$. Looking at the start of these words (first letter, first two letters, first three letters, etc.) we notice that the starts all contain at least as many U's as R's.

Problem 8.3 Let $A = \{1, 2, 3, 4\}$ and $B = \{0, 1\}$.

(a) How many functions from A to B are there?

Answer

16

Copyright © ARETEEM INSTITUTE. All rights reserved.

Solution

For each of the four elements in A we choose either 0 or 1 from B (two choices). Therefore there are $2^4 = 16$ functions in total.

(b) How many subsets of A are there?

Answer

16

Solution

For each of the four elements in A we choose either to include it or exclude it from the subset. Therefore there are $2^4 = 16$ subsets in total.

(c) Explain your answers in (a) and (b) using a bijection.

Solution

We want to associate each function in (a) to a subset from (b). Given a function f from (a), associate it with the subset containing the elements x from A with $f(x) = 1$. For example, if we have a function $f(x)$ such that $f(1) = f(3) = f(4) = 1$ and $f(2) = 0$ the subset it is associated with is $\{1, 3, 4\}$.

Problem 8.4 Let $A = \{1, 2, 3, 4, 5\}$.

(a) Give an example of a bijection between subsets of A of size two and subsets of A of size three.

Solution

Let B be a subset of A of size 2. Map B to $A \setminus B$. For example, $\{1, 4\} \mapsto \{2, 3, 5\}$. It is not hard to see this is a bijection.

(b) How many such bijections are there in part (a)?

Answer

10!

Copyright © ARETEEM INSTITUTE. All rights reserved.

Solution

Note that each set has size

$$10 = \binom{5}{2} = \binom{5}{3}.$$

If we fix the subsets of size 2, any rearrangement and pairing of the subsets of size 3 gives a bijection, so there are 10! bijections.

Problem 8.5 Let $A = \{1, 2, 3, 4\}$.

Find a bijection between subsets of A of size 2 and two numbers chosen (repetition allowed) from $\{1, 2, 3\}$.

Solution

Suppose the subset of A is $\{a, b\}$ with $a < b$. Map this to the two numbers $a, b-1$. Convince yourself that this is a bijection.

Problem 8.6 Stars and Bars

(a) You want to arrange 6 stars ($*$'s) and 3 bars ($|$'s) in a line. How many ways are there for you to do so?

Answer

84

Solution

In total we have $6 + 3 = 9$ symbols. As the symbols of each type are all identical, there are $\binom{9}{6} = \binom{9}{3} = \dfrac{9!}{6! \cdot 3!} = 84$ ways to arrange them.

(b) How many ways are there to choose non-negative integers a, b, c, d such that $a + b + c + d = 6$? Explain how your answer in part (a) can help.

Answer

84

Copyright © Areteem Institute. All rights reserved.

Solution

Represent the number 6 using the six stars as $******$. To break the 6 into the sum of 4 integers, we divide the stars $******$ into 4 groups using $4 - 1 = 3$ bars. For example $1 + 2 + 2 + 1$ is represented by $*|**|**|*$. Therefore the number of ways to choose the integers a, b, c, d is the same as the number of ways to arrange 6 stars and 3 bars, which we know was 84 from (a).

Problem 8.7 Only Positive Numbers

(a) Consider 5 plus signs in a row: $+ + + + +$. How many ways are there to choose 3 plus signs to remove? For example, if you remove the first, second, and fourth you are left with $\_ \_ + \_ +$.

Answer

10

Solution

There are 5 plus signs in total, so we just need to choose 3 to remove (the order doesn't matter). This can be done in $\binom{5}{3} = 10$ ways.

(b) How many ways are there to choose positive integers a, b, c such that $a + b + c = 6$? Explain how your answer in part (a) can help.

Answer

10

Solution

Representing the 6 as $1 + 1 + 1 + 1 + 1 + 1$ note we have 5 plus signs in total. If we want 3 numbers (a, b, c) in total, we need to remove three of the plus signs and regroup the ones. For example, if we remove the first 3 plus signs we would get the sum $4 + 1 + 1$. Therefore the number of ways to choose a, b, c is the same as the number of ways to remove 3 of the plus signs, which we know from (a) can be done in 10 ways.

Problem 8.8 How many ways are there to choose non-negative integers a, b, c, d such that $a + b + c + d = 6$ with $a = 1$ and $b \geq 2$?

Copyright © ARETEEM INSTITUTE. All rights reserved.

Answer

10

Solution

We can again use stars and bars. We have a total of 6 stars and $4 - 1 = 3$ bars. However, for a to equal 1 we must set aside a star and bar so that our arrangement can start with $*|$. For $b \geq 2$ we set aside 2 stars to insert between the second and third bars. We are then left with 2 bars and 3 bars, which can be arranged in $\binom{5}{3} = 10$ ways.

Problem 8.9 Suppose 5 people get in an elevator on Floor 0. The people leave the elevator somewhere between (inclusive) Floors 1 and Floor 5.

(a) If we only care about how many people get of at each floor, how many ways can the people get off?

Answer

126

Solution

If we let a, b, c, d, e denote the number of people who get off on Floors $1, 2, 3, 4, 5$, we need $a + b + c + d + e = 5$. As it is possible for no-one to get off on any specific floor, we use the non-negative version of stars and bars for a total of $\binom{5+5-1}{5} = \binom{9}{5} = 126$ outcomes.

(b) If we only care about what collection of floors the elevator stops on, how many different collections are there?

Answer

$2^5 - 1 = 31$

Solution

Note that we have enough people that it is possible for every floor to be stopped on. Hence, it is possible for the elevator to stop or not (2 choices) on each floor. However, it must stop on at least one floor, so we subtract 1.

Copyright © ARETEEM INSTITUTE. All rights reserved.

Problem 8.10 You line up 11 cards in a row. 8 of the cards are black and identical. The other 3 cards are red and numbered 1, 2, and 3. How many different ways to line up the 11 cards are there if each of the red cards is separated by at least 2 black cards?

Answer

210

Solution

First arrange the red cards in one of $3! = 6$ ways. We then must place the 8 black cards in one of the four spaces as in the diagram below:

$$\_\,R\,\_\,R\,\_\,R\,\_$$

Set aside $2 \cdot 2 = 4$ black cards to separate the three red cards by at least two cards. We are left with $8 - 4 = 4$ identical black cards that can be placed in any of the 4 slots. For these we can use the non-negative version of stars and bars to see there are $\binom{4+4-1}{4} = \binom{7}{4} = 35$ ways to arrange the black cards. This gives $6 \cdot 35 = 210$ total outcomes.

Copyright © ARETEEM INSTITUTE. All rights reserved.

9 Solutions to Chapter 9 Examples

Problem 9.1 You roll 2 four-sided dice. Let A be the event that the first die is a 4, and B the event that the sum of the two rolls is 6.

(a) What is $P(A)$?

Answer

$\dfrac{1}{4}$

Solution

We have $A = \{(4,1),(4,2),(4,3),(4,4)\}$ so $n(A) = 4$. Since $n(\Omega) = 4^2 = 16$, we have $P(A) = \dfrac{4}{16} = \dfrac{1}{4}$.

(b) What is $P(B)$?

Answer

$\dfrac{3}{16}$

Solution

We have $B = \{(2,4),(3,3),(4,2)\}$ so $n(B) = 3$. Since $n(\Omega) = 4^2 = 16$, we have $P(B) = \dfrac{3}{16}$.

(c) What is $P(A \cap B)$?

Answer

$\dfrac{1}{16}$

Solution

The only outcome in $A \cap B$ is $(4,2)$ so $n(A \cap B) = 1$. Since $n(\Omega) = 4^2 = 16$, we have $P(A \cap B) = \dfrac{1}{16}$.

Problem 9.2 Suppose you flip a fair coin 6 times.

Copyright © ARETEEM INSTITUTE. All rights reserved.

(a) Find the probability of exactly 4 heads.

Answer

$\dfrac{15}{64}$

Solution

Ω is the set of all words of length 6 using the letters H, T, so $n(\Omega) = 2^6$. If A is the event that we get exactly 5 heads,

$$n(A) = \binom{6}{4} = 15,$$

so

$$P(A) = \frac{\binom{6}{4}}{2^6} = \frac{15}{64}.$$

(b) Find the probability of at least one tails.

Answer

$\dfrac{63}{64}$

Solution

Note there is only 1 way to get 0 tails. If A is the event that we get at least one tails, then $n(A) = 64 - 1 = 63$ (using complementary counting). Hence the probability is $\dfrac{63}{64}$.

Problem 9.3 A dealer starts with only the 4 aces (one of each suit) from a deck of cards. They deal you 2 of the cards. Let A be the event that you get one heart and one diamond. Let B be the event that you get a spade.

(a) Assume the cards are dealt to you in order (that is, a first card and a second card). Find $P(A)$ and $P(B)$. Hint: You may want to write out a sample space.

Answer

$P(A) = \dfrac{1}{6}, P(B) = \dfrac{1}{2}$

Copyright © ARETEEM INSTITUTE. All rights reserved.

Solution

The sample space Ω has size $4 \cdot 3 = 12$. $n(A) = 2$ (the 2 orders of the cards) and $n(B) = 2 \cdot 3$ (there are 3 choices for which other card to get, and then 2 orders).

$$P(A) = \frac{2}{12} = \frac{1}{6}, P(B) = \frac{6}{12} = \frac{1}{2}$$

(b) Assume the cards are not dealt in order (that is, you are dealt two cards at once). Find $P(A)$ and $P(B)$. Hint: You may want to write out a sample space.

Answer

$$P(A) = \frac{1}{6}, P(B) = \frac{1}{2}$$

Solution

Now the the sample space Ω has size $\binom{4}{2} = 6$. $n(A)$ is now 1, and $n(B) = 3$. (We no longer have orders to worry about.)

$$P(A) = \frac{1}{6}, P(B) = \frac{3}{6} = \frac{1}{2}$$

(c) Compare your answers in parts (a) and (b). Can you explain the outcome?

Solution

They are the same. When we are picking without replacement, as long as we are consistent about order or not order, we will get the same answer for probability.

Problem 9.4 You have 5 red, numbered 1 through 5, and 8 green balls, numbered 1 through 8. You pick 5 without replacing the balls. For each of the events below, find the probability. Note: It is best to think of all 5 balls being picked at once.

(a) What is the probability you get 3 green and 2 red balls?

Answer

$$\frac{560}{1287}$$

Copyright © ARETEEM INSTITUTE. All rights reserved.

Solution

The total number of outcomes in the sample space is

$$\binom{13}{5} = 1287$$

since we are interested in choosing 5 balls from a group of 13 balls. Of the 5 balls, there are

$$\binom{5}{2} = 10$$

ways to decide which red balls to get and

$$\binom{8}{3} = 56$$

ways to decide which green balls to get. Therefore, the probability is

$$\frac{\binom{5}{2}\binom{8}{3}}{\binom{13}{5}} = \frac{56 \times 10}{1287} = \frac{560}{1287}$$

(b) What is the probability that all the balls you pick are the same color?

Answer

$$\frac{19}{429}$$

Solution

The total number of outcomes in the sample space is

$$\binom{13}{5} = 1287$$

since we are interested in choosing 5 balls from a group of 13 balls. We have two cases to consider.

The first case is when all 5 balls are red. There are

$$\binom{5}{5} = 1$$

Copyright © ARETEEM INSTITUTE. All rights reserved.

way of choosing 5 balls to be red.

The second case is when all 5 balls are green. There are

$$\binom{8}{5} = 56$$

way of choosing 5 balls to be green. Therefore, the desired probability is

$$\frac{\binom{5}{5} + \binom{8}{5}}{\binom{13}{5}} = \frac{57}{1287} = \frac{19}{429}.$$

Problem 9.5 You have 3 red balls, numbered 1 through 3, and 4 green balls, numbered 1 through 4. You pick 5 balls, one by one, replacing the ball after each pick. (Thus you can pick the same ball more than once.)

(a) How many outcomes are there where you pick 2 red and 3 green balls?

Answer

5760

Solution

We are picking with replacement, so it is possible to pick the same ball more than once. For each red ball we have 3 choices and for each green ball we have 4 choices. Since we also need to decide which of the five picks are red (the other three are green) there are

$$\binom{5}{2} \cdot 3^2 \cdot 4^3 = 10 \cdot 9 \cdot 64 = 5760$$

ways to pick 2 red and 3 green balls.

(b) What is the probability of picking 2 red and 3 green balls?

Answer

$$\frac{5760}{16807}$$

Solution

We know from part (a) that there are 5760 outcomes we want. There are 7 balls in all, so

Copyright © ARETEEM INSTITUTE. All rights reserved.

picking with replacement there are $7^5 = 16807$ outcomes in total. Hence the probability is $\dfrac{5760}{16807}$.

Problem 9.6 Assuming only the axioms Pr1 - Pr3:

Pr1. $P(A) \geq 0$.
Pr2. $P(\Omega) = 1$.
Pr3. If A and B are *disjoint* (that is, $A \cap B = \emptyset$), then $P(A \cup B) = P(A) + P(B)$.

Prove:

(a) Pr6: $P(A^c) = 1 - P(A)$.

Solution

For any event A, A and A^c are disjoint and $A \cup A^c = 1$. Using Pr1 we have $1 = P(\Omega)$. Pr3 tells us that $P(\Omega) = P(A) + P(A^c)$. Hence we have

$$1 = P(\Omega) = P(A) + P(A^c) \Rightarrow P(A^c) = 1 - P(A)$$

which is Pr6 as needed.

(b) Pr4: $P(\emptyset) = 0$.

Solution

Note that $\Omega^c = \emptyset$. Hence using Pr2 and Pr6 (which we proved in (a)), we know $P(\emptyset) = P(\Omega^c) = 1 - P(\Omega) = 1 - 1 = 0$ as needed.

Problem 9.7 Emily builds an unfair 6-sided die. The probability of rolling a 1 is the same as rolling a 3. Rolling a 4 is twice as likely as rolling a 1 and rolling a 5 is 10% more likely than rolling a 1. If the probability of rolling a 2 is 0.2 and rolling a 6 is 0.1, find the probability of rolling each of the six sides of the die.

Answer

$P(1) = 0.12, P(2) = 0.2, P(3) = 0.12, P(4) = 0.24, P(5) = 0.22, P(6) = 0.1$

Solution

Let $P(1) = x$. Then

$$P(3) = x, P(4) = 2 \cdot P(1) = 2x, \text{ and } P(5) = P(1) + 0.1 = x + 0.1.$$

Copyright © Areteem Institute. All rights reserved.

We are given $P(2) = 0.2$ and $P(6) = 0.1$. Hence

$$1 = P(1) + P(2) + P(3) + P(4) + P(5) + P(6)$$
$$= x + 0.2 + x + 2x + (x + 0.1) + 0.1$$
$$= 5x + 0.4$$

Hence solving $5x + 0.4 = 1$ we have $x = 0.12$. In turn this gives

$$P(1) = P(3) = 0.12, P(4) = 2 \cdot 0.12 = 0.24, \text{ and } P(5) = 0.12 + 0.1 = 0.22,$$

as our missing probabilities.

Problem 9.8 Probability Venn Diagram Practice

(a) If $P(A) = 0.5$, $P(B) = 0.7$, what are the maximum and minimum possible values of $P(A \cap B)$?

Answer

Max: 0.5, Min: 0.2

Solution

If A is contained in B, then $P(A \cap B) = P(A) = .6$ (and this is the maximum). The minimum overlap will occur when $P(A \cup B) = 1$. In this case $P(A \cap B) = 0.5 + 0.7 - 1 = 0.2$.

(b) Suppose $P(A) = 0.5$, $P(B) = 0.7$, $P(A^c \cap B^c) = 0.2$. Find $P(A \cap B^c)$ and $P(B \cap A^c)$.

Answer

$P(A \cap B^c) = 0.2$, $P(B \cap A^c) = 0.3$

Solution

$1 = P(A^c \cap B^c) + P(A) + P(B \cap A^c)$ so $P(B \cap A^c) = 0.3$. From here it follows that $P(A \cap B) = 0.4$ and thus $P(A \cap B^c) = 0.1$.

Problem 9.9 Suppose you flip a fair coin 6 times.

(a) Find the probability of no two heads in a row and no two tails in a row.

Copyright © ARETEEM INSTITUTE. All rights reserved.

Answer

$\dfrac{1}{32}$

Solution

Note the only possibilities are $HTHTHT, THTHTH$.

(b) Find the probability you get more heads than tails.

Answer

$\dfrac{11}{32}$

Solution

Since the coin is fair, using symmetry the probability you get more heads than tails is the same as the probability you get more tails than heads. However, it is possible to get the same number of heads as tails because we have an even number of flips.

The probability we get 3 heads and 3 tails is

$$\frac{\binom{6}{3}}{2^6} = \frac{20}{64} = \frac{5}{16}.$$

This gives

$$\left[1 - \frac{5}{16}\right] \div 2 = \frac{11}{16} \div 2 = \frac{11}{32},$$

as the probability you get more heads than tails.

Problem 9.10 Suppose you roll a fair die 4 times. What is the probability the sum of all 4 rolls is 10?

Answer

$\dfrac{5}{81}$

Solution

There are $6^4 = 1296$ total outcomes for rolling the die 4 times.

If we let a, b, c, d denote the rolls, we want $a + b + c + d = 10$. Using the positive version

Copyright © ARETEEM INSTITUTE. All rights reserved.

of stars and bars there are $\binom{9}{3} = 84$ outcomes with $a, b, c, d \geq 1$. However, some of these outcomes, such as $1 + 1 + 1 + 7 = 10$ do not correspond to possible dice rolls. In fact, the 4 rearrangements of $1 + 1 + 1 + 7 = 10$ are the only outcomes that do not work. Hence there are 80 ways for the sum of all 4 rolls to equal 10 and hence the probability is $\dfrac{80}{1296} = \dfrac{5}{81}$.

Copyright © ARETEEM INSTITUTE. All rights reserved.

www.ingramcontent.com/pod-product-compliance
Lightning Source LLC
Chambersburg PA
CBHW081505200326
41518CB00015B/2387